A STRATEGIC ATLAS

The policy of a state lies in its geography.
—NAPOLEON

A STRATEGIC ATLAS

Comparative Geopolitics of the World's Powers

SECOND EDITION, REVISED AND UPDATED

Gérard Chaliand and Jean-Pierre Rageau

Translated from the French by Tony Berrett

Maps by Catherine Petit

HARPER & ROW, PUBLISHERS, New York
Cambridge, Philadelphia, San Francisco, London
Mexico City, São Paulo, Sydney

THIS ATLAS IS DEDICATED

- to the British geopolitical analyst Halford J. Mackinder (1861–1947);

- to the theorist of sea power, the American Alfred T. Mahan (1840–1914);

- to the pioneer of geopolitics, the German Friedrich Ratzel (1844–1904);

- and to the French geographer Pierre Vidal de la Blache (1845–1918)

This work was originally published in France under the title *Atlas Stratégique,* © Librairie Arthème Fayard, 1983.

FIRST EDITION

ISBN: 0-06-015387-3

ISBN: 0-06-091220-0 (pbk.)

LIBRARY OF CONGRESS CATALOG CARD NUMBER: 84-48143

86 87 88 10 9 8 7 6 5 4 3 2

PREFACE

This atlas is an innovation. No strategic atlas exists in English, French, or German—nor, so far as we are aware, in any other language. For this is not a mapping of past or future battles, or a graphic depiction of opposing military forces. Strategy, like politics, embraces war, but is more than war. What is provided here is truly a geopolitics of the relations of force in the contemporary world. In order to portray as accurately as possible realities that are at once multifaceted, complex, and sometimes impossible to represent at all (how, for example, to measure determination, which is of course a key factor in conflict situations? Can one predict surprise?), we have sought to be as global as possible.

From the outset, our perspective on our planet breaks with the Mercator projection, with its horizontal and almost pre-Galilean world, in which the land masses appear to cover a larger area than the seas. The modern perspective should give a more faithful picture of a globe on which the poles, the axes of the world, are, at least at the present time in the case of the Arctic, a decisive area that does not appear on the usual maps. We have therefore favored a multiplicity of projections, and these have been selected solely for the purposes of our demonstration, and to show every aspect of the world.

Thus, geographically our atlas gives considerable space to the oceans, which cover the larger part of the planet, on which sea power is engaged, mastery of which has ensured Anglo-American hegemony for almost two centuries: the Arctic Ocean whose frozen surfaces do not prevent the passage of nuclear submarines underneath; the Pacific, whose strong points are all controlled by the Anglo-Americans and the French; the Atlantic, both North and South; and the Indian Ocean, currently a delicate area where more than anywhere else today's rivalries manifest themselves.

We have also sought to show, in addition to the usual views and projections used by Europeans and Americans, the world as it is seen by the Chinese, the Soviets, and the Arab Muslims. We have sought to show the great cultural and religious domains, which are the matrices of world views, and in this way to complement our strategic approach with the historical background that shapes collective behaviors and that may determine external decisions to fight, or provoke imbalances within a society.

No attempt has ever been made, for example, to compile a map of the major traditional enmities that have determinied conflicts in a given historical epoch. How, for example, can we understand Poland without alluding to its dual rejection of the Russians and the Germans? However, our atlas is essentially centered on the at least militarily bipolarized world that emerged from World War II. It shows the present world strategic situation while at the same time indicating the changes that have occurred since the beginnings of the Cold War, both in the area of crises and in conventional conflicts or guerrilla activities.

This strategic atlas also looks at a dimension that is not treated in the usual works: states' perceptions of their security—not only those of the world powers (the United States or the Soviet Union), but also those of the lesser known regional powers such as Saudi Arabia, India, South Africa, Brazil, Japan, and Israel. It is, for example, not sufficiently appreciated that the security perceptions of a state such as Saudi Arabia are a function of its hostility not only to the Soviet Union and Israel, but also to the Soviet presence in Afghanistan, Ethiopia (an obstacle to Arab control of the Red Sea), and the Democratic Republic of Yemen on its southern flank; of its rejection of a regional hegemony by Shiite Iran; and finally of its own military and demographic weakness, half its working population being foreigners.

We have also tried to present a multiform world with very different perceptions*. Of course, our work includes a number of more conventional maps concerning agricultural, industrial, mineral, and energy resources, as well as demographic data and information making it possible to grasp the nature of North–South relations. Wherever it seemed useful, we have provided statistical projections up to the end of the century. Finally, our atlas includes a strictly military section, partly centered on nuclear questions. In short, our conception of strategy attempts to embrace the totality of human, material, and cultural factors that make up a global balance of forces.

As such, our atlas does not claim to be free of all shortcomings. But we are nevertheless certain that we are taking a novel approach to revealing the political realities of the contemporary world.

We would like to thank General Pierre Gallois, who looked at our military maps and generously provided two that he had drawn himself; Catherine Petit, our cartographer, for her work; and Claude Durand, who was from the beginning convinced of the novelty of our project and took on the risks involved.

Gérard Chaliand
Jean-Pierre Rageau

*Perhaps we have been helped in this by the fact that we both studied at the Ecole Nationale des Langues Orientales, one of the very few institutions where it was possible in the 1950s to acquire a view of the world that was not Western-centered.

CONTENTS

Polar Projection

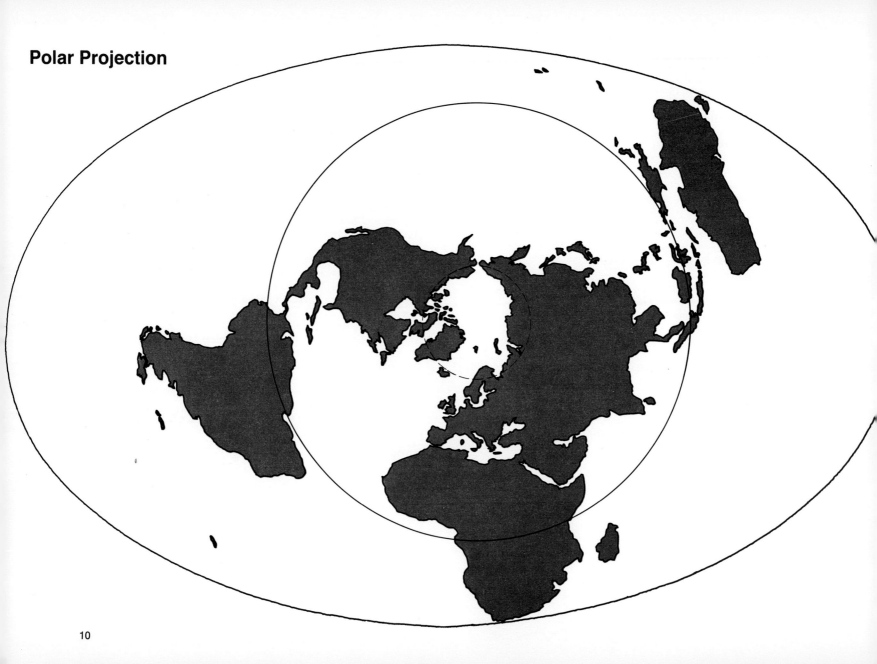

Circular Projection

The earth is a sphere. Any representation of it on a
two-dimensional plane is obtained by a projection.
For example, the map on the left is a polar projection. For convenience, most maps and atlases use
plane projections, which ignore the spherical nature
of the globe. But this convention falsifies strategic
perception as soon as it goes beyond the regional
level. The circular projection on the right is modi-
fied, and hence misleading, but serves to illustrate
our point.

A Militarily Bipolar World

China, which is militarily hostile to the USSR, cannot be included among the West's allies. (For the African states linked to France by assistance agreements, see the section on Africa.)

USA

U.S. allies

China

USSR allies

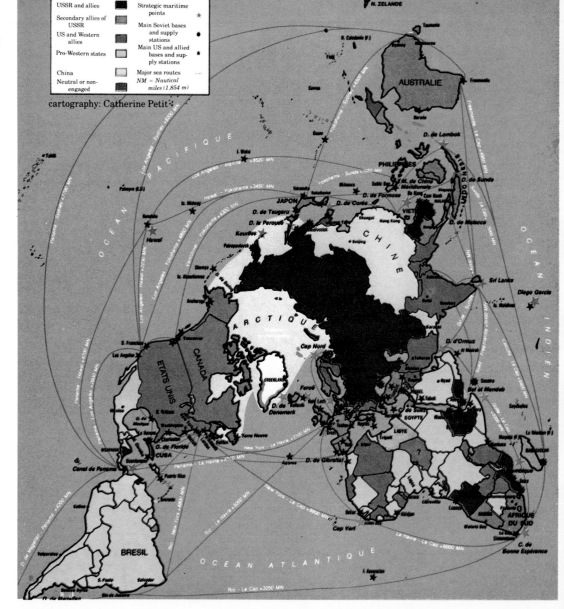

cartography: Catherine Petit

USSR and allies	Strategic maritime points
Secondary allies of USSR	Main Soviet bases and supply stations
US and Western allies	Main US and allied bases and supply stations
Pro-Western states	Major sea routes
China	NM = Nautical miles (1.854 m)
Neutral or non-engaged	

VIEWS OF THE WORLD

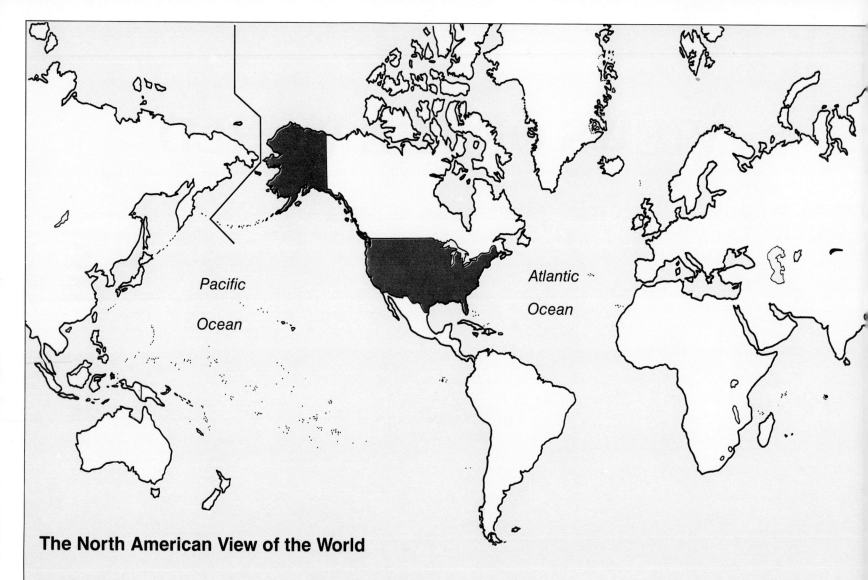

Pacific

Ocean

Atlantic

Ocean

The North American View of the World

The Old World, Asia, and Africa on the other side of the oceans.

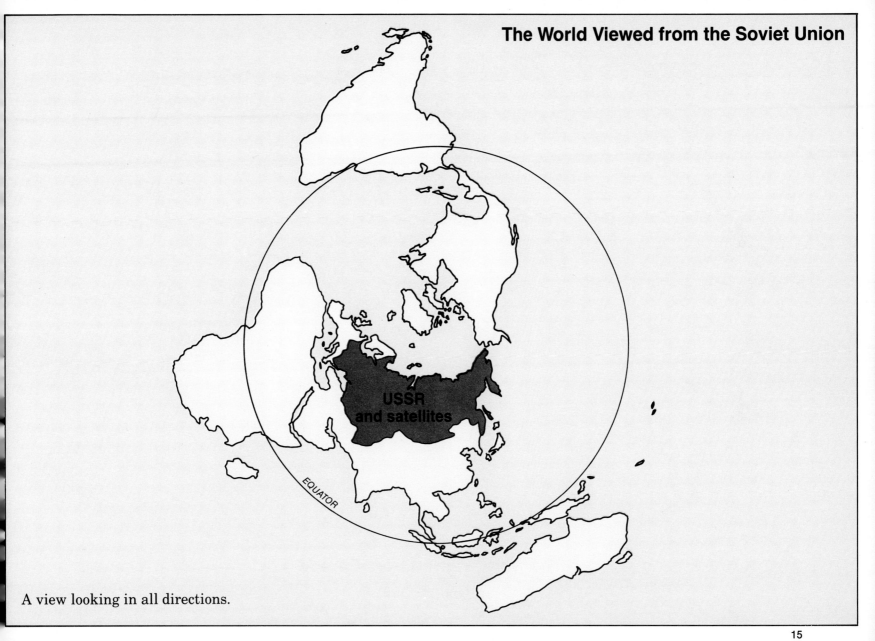

USSR
and satellites

EQUATOR

A view looking in all directions.

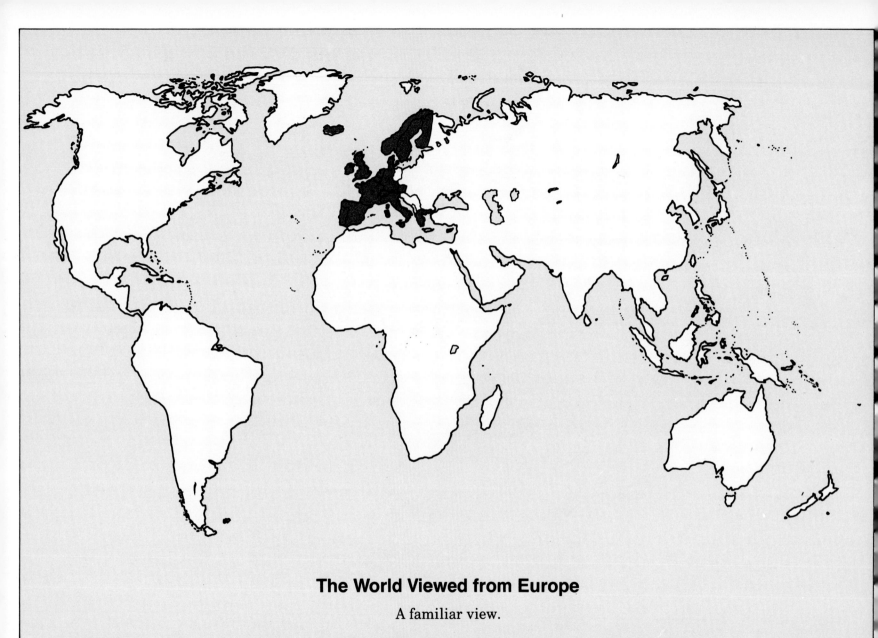

The World Viewed from Europe

A familiar view.

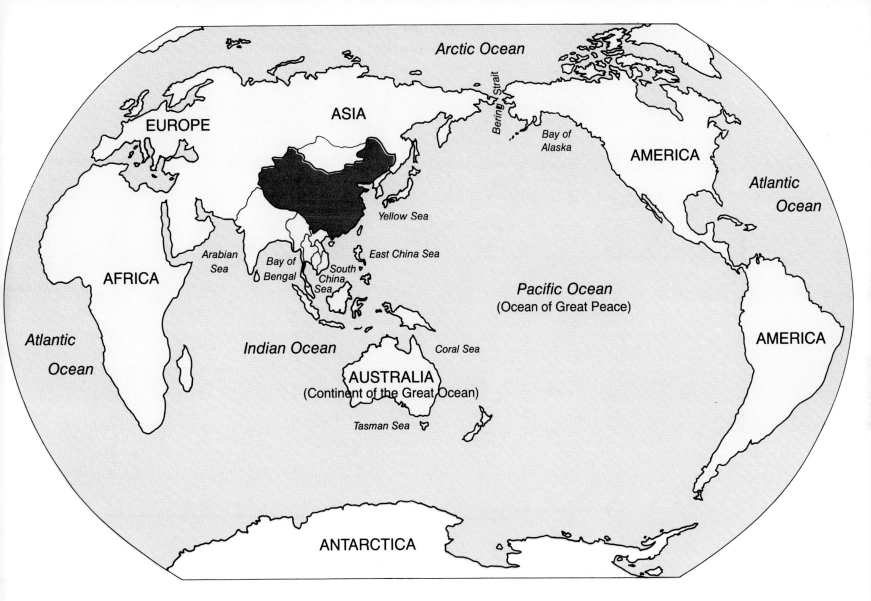

Arctic Ocean

ASIA

EUROPE

Bering Strait

Bay of
Alaska

AMERICA

Atlantic
Ocean

AMERICA

AFRICA

Yellow Sea

Arabian
Sea

Bay of
Bengal

South
China
Sea

East China Sea

Pacific Ocean
(Ocean of Great Peace)

Atlantic

Ocean

Indian Ocean

Coral Sea

AUSTRALIA
(Continent of the Great Ocean)

AMERICA

Tasman Sea

ANTARCTICA

The World Viewed from China

From the *Contemporary Atlas of the People's Republic of China.*

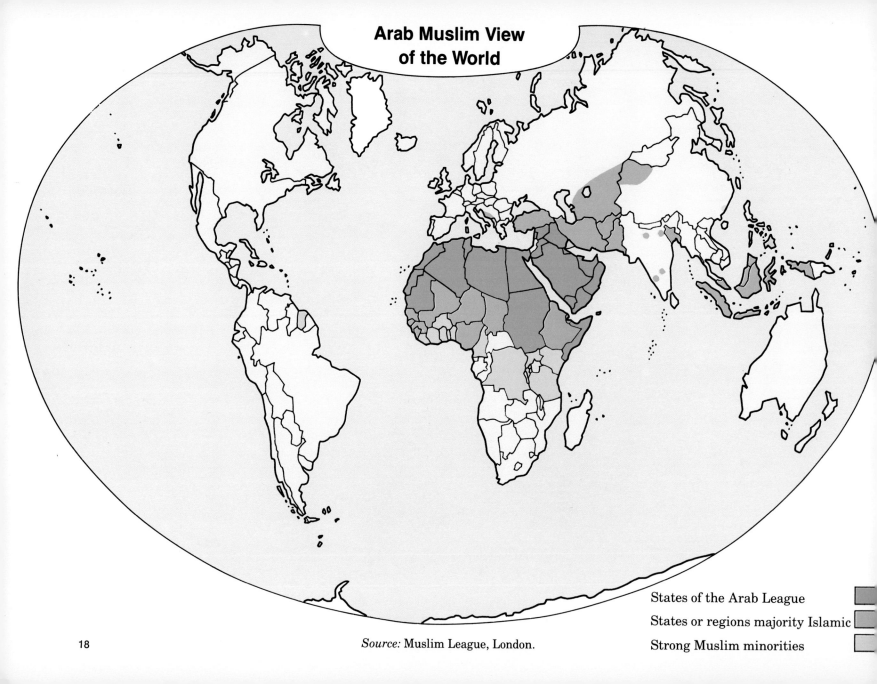

Arab Muslim View of the World

States of the Arab League

States or regions majority Islamic

Strong Muslim minorities

Source: Muslim League, London.

THE GEOPOLITICIANS

The Geopoliticians

The German geographer Friedrich Ratzel, author of *Politische Geographie* (1897), developed a number of basic concepts, particularly concerning *space,* that have inspired geopoliticians. It was the British writer H. Mackinder who in 1904 proposed the notion that the continental part of Eurasia, by virtue of its land mass, forms the world Heartland. According to Mackinder (maps pp. 21 and 22), who several times revised his geographical delimitation of the heartland (in 1919 and 1943), the power that controls this land mass—once potentially Germany, now the Soviet Union—threatens the sea powers—once Great Britain, now the United States—that control the World Island—that is, our planet.

The factors that Mackinder came to include as his thinking developed were communications (including aviation), population, and industrialization. In 1943, he repudiated his 1919 theory (the state that controls the Heartland will dominate the World Island).

The American Mahan, a geopolitician before the word was invented, put forward as early as 1900 (in *The Problem of Asia and Its Effect upon International Politics*) the idea that the world hegemony of sea powers can be maintained by control of a series of bases around the Eurasian continent. This view foreshadowed Mackinder's concept of the World Island, but it led to the opposite strategic conclusions: Sea powers dominate land powers by hemming them in. Therein lies the seed of the theory of containment born of the Cold War.

Geopolitical conceptions were systematized by the Swede Rudolf Kjellen and then adopted by the German geopoliticians, especially Karl Haushofer (1869–1946). German geopolitics developed in three directions: the concept of space *(Raum)* brought out by Ratzel, meaning the need for a great power to have space available to it; the concept of a World Island enunciated by Mackinder, implying sea power; and the North–South combination of continents put forward by Haushofer (map p. 24). This latter conception is to be found today, for example, in the Eurafrican policy of Western Europe.

The American N. J. Spykman (map p. 23) followed Mackinder and adapted his concepts to the circumstances of the 1930s. He argued that only an Anglo-American (sea power) and Russian (land power) alliance could prevent Germany from controlling the Eurasian coastal regions and thus achieving world domination. But he rejected some of Mackinder's strategic conclusions concerning the importance of controlling the Heartland by giving greater importance to control of the Rimland.

Although it is sometimes excessively systematic, the geopolitical approach is stimulating; but it is so only if there is no lapse into geographical determinism and if all factors in the balance are taken into account. In the map on p. 25 we sketch our own approach along these lines, one more in conformity with present-day realities.

The World according to Mackinder (1904)

Control of the Heartland, the
Eurasian continental land mass,
constitutes a potential threat for
sea powers.

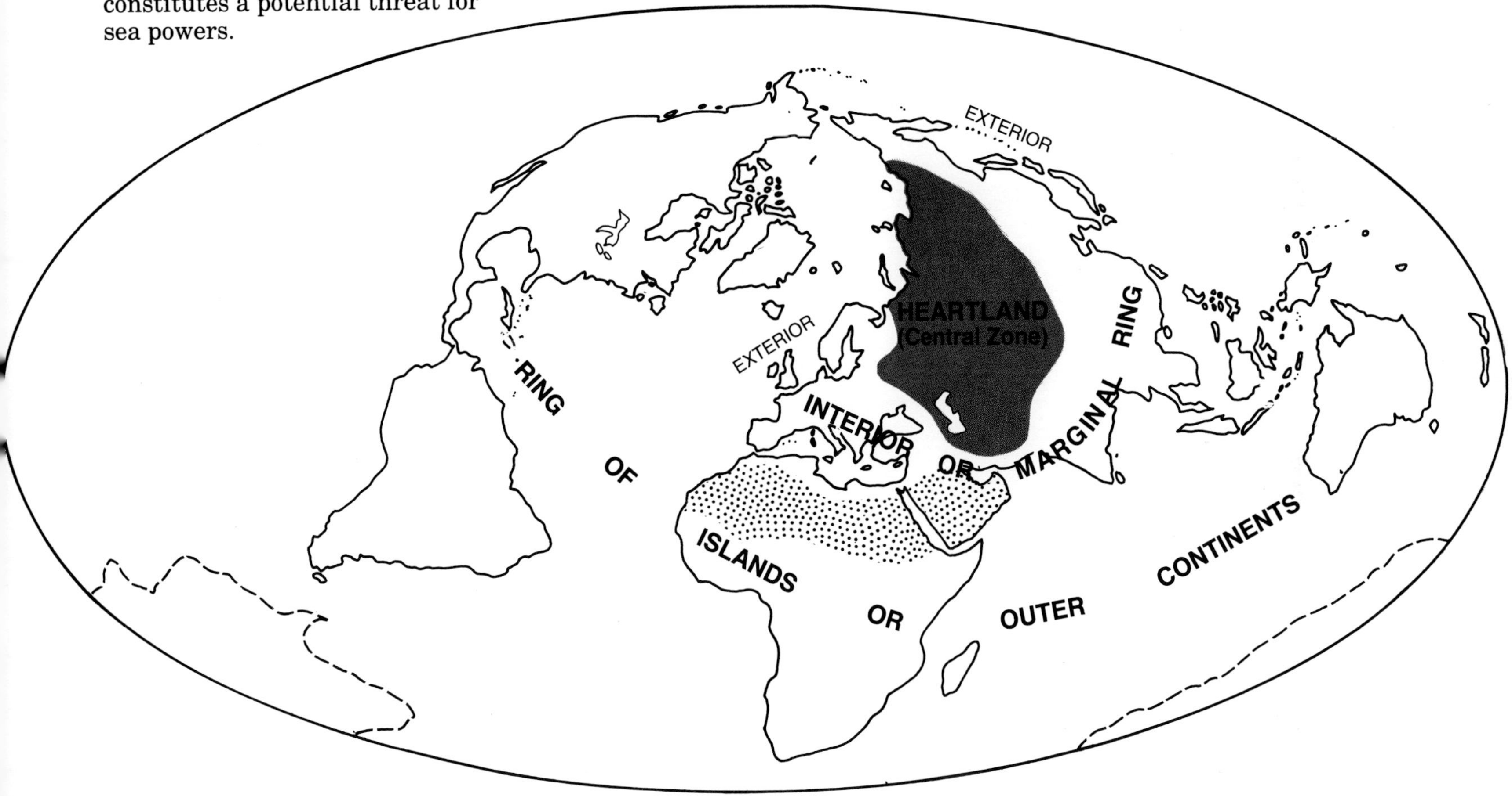

EXTERIOR

EXTERIOR

HEARTLAND
(Central Zone)

RING

OF

ISLANDS

OR

INTERIOR

OR

MARGINAL

RING

OUTER

CONTINENTS

The World according to Mackinder (1943)

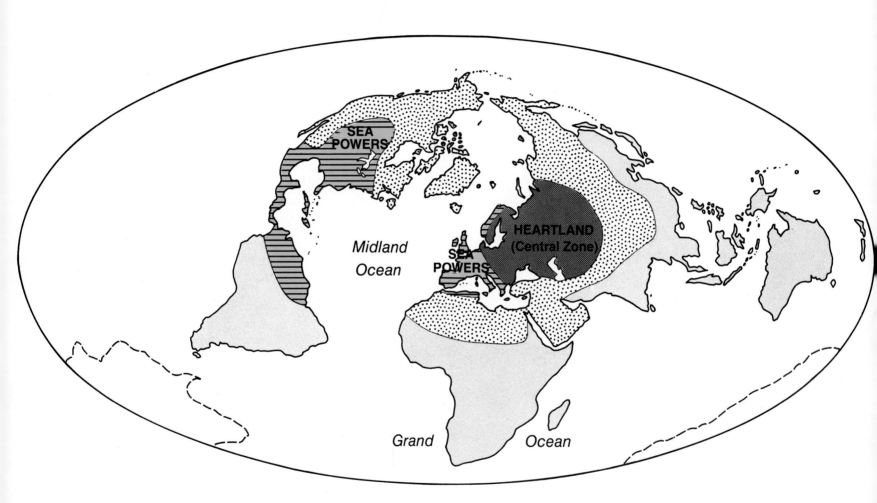

SEA POWERS

SEA POWERS

HEARTLAND (Central Zone)

Midland Ocean

Grand Ocean

Spykman and the Importance of the Rimland

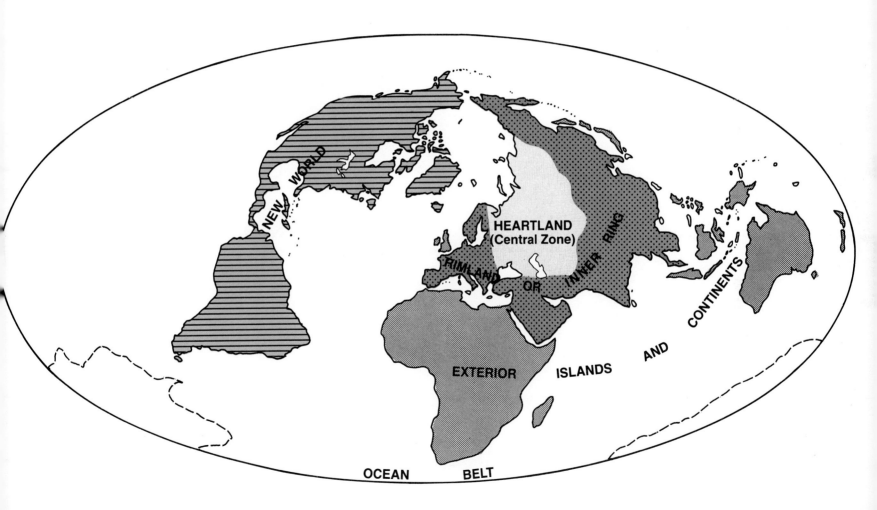

Haushofer and the North–South Combination

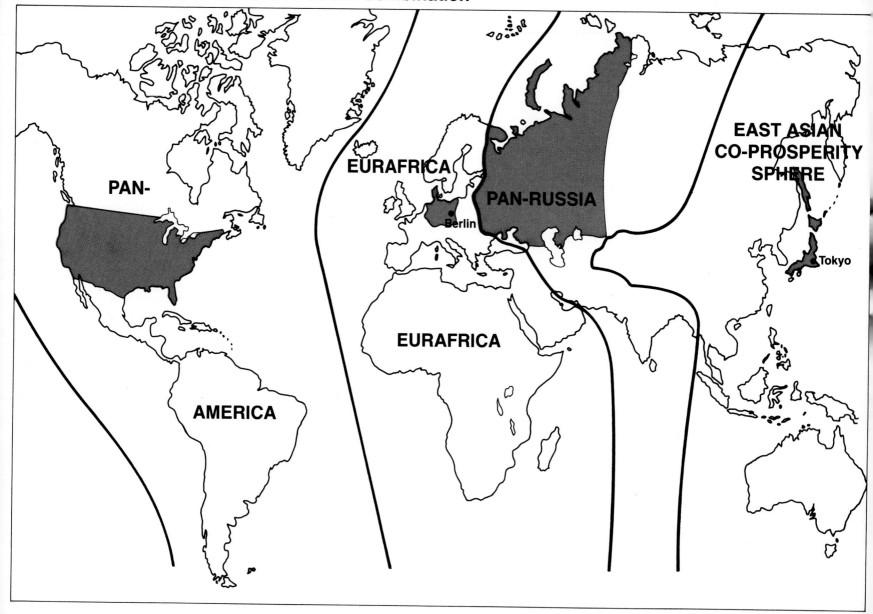

Sketch of a Contemporary Geopolitics

Taking into account inevitable developments and changes, we see that geopoliticians before World War II accurately assessed these basic factors:
- *Heartland*
- *Sea power*
- *Rimland*

We also see the appearance since 1945, with the independence of new states in Asia and Africa, of a fragile and unstable intertropical ring:
- *Ring of underdevelopment and poverty*

Gradual emergence of a *developed southern ring* is linked to sea power.

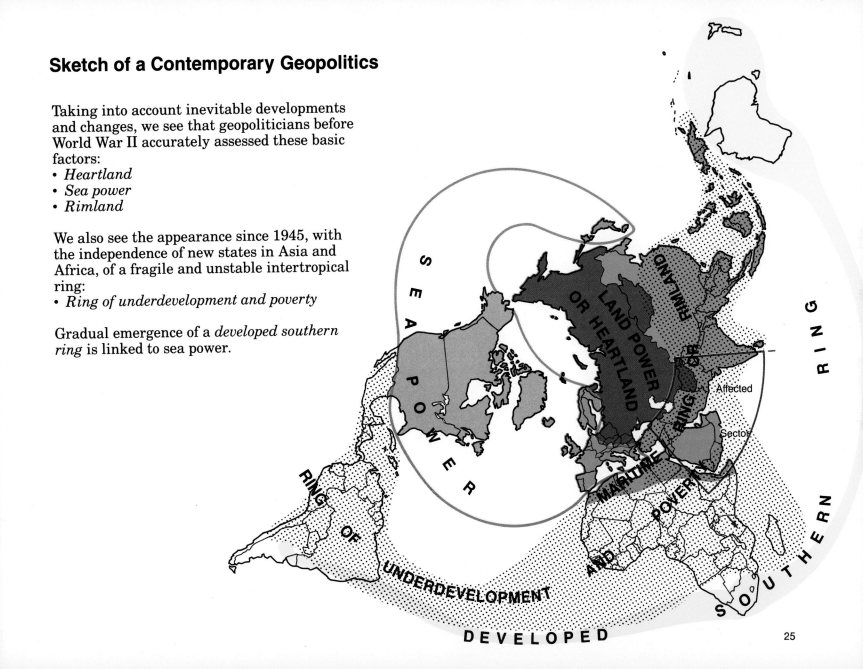

25

CULTURAL FACTORS

The Great Cultural Domains

The cultural area of which Europe is the center has undergone considerable expansion. So has that of Islam which, in Southeast Asia, covers regions that were formerly Hindu, while at the same time continuing its expansion in Africa.*

*Madagascar is attached to Africa for the sake of convenience, but ethnically does not belong to the African world.

European domain

Chinese domain

Russo-Soviet domain

Islamic domain

Black African domain

Hindu domain

Latin American domain

South Africa (mixed)

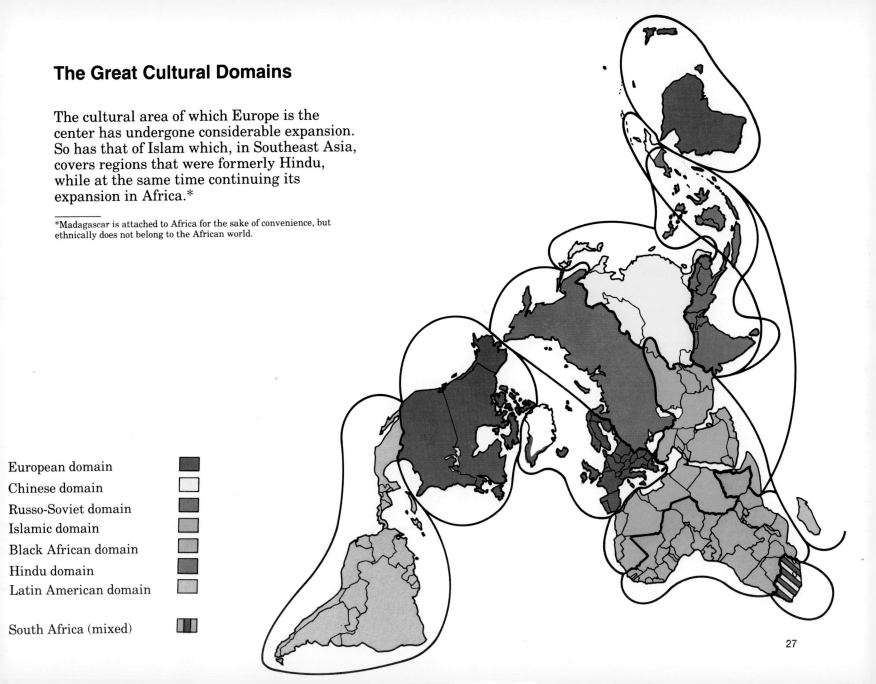

The Great Religions

Religions were formerly the major source of
identity, and since nationalism, rather than
superseding religion, is often superimposed on
it, religions continue to play a leading role.

Notes:
- Oriental Christians: principally Orthodox
- China: superimposition and
  interdependence of Confucianism,
  Buddhism, Taoism
- Japan: superimposition of Shintoism and
  Buddhism
- Gray areas: animism and others

Catholicism

Protestantism

Oriental Christians

Islam (Sunni)

Islam (Shiite)

Buddhism

Hinduism

Chinese syncretism

Japanese syncretism

Catholic minorities   +

Judaism   ✳

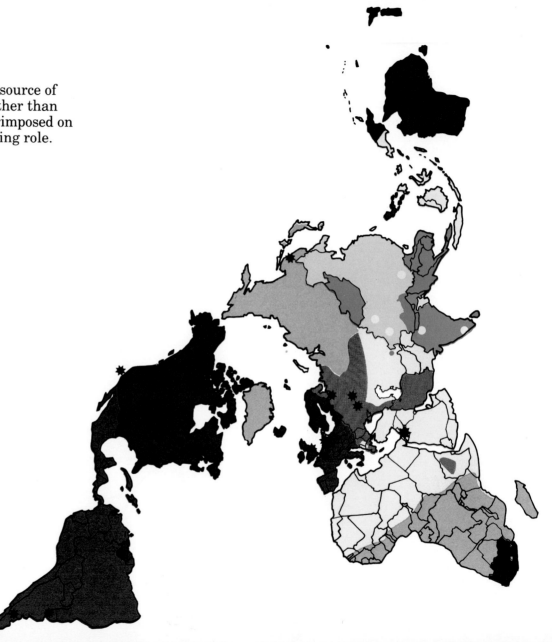

28

Imperial Languages of the World

The criterion used to define an imperial language is the combination of its numerical importance *and* its geographic diffusion. In this respect Japanese is not imperial, and German since 1945 has lost its preponderance in Eastern Europe. Chinese (Peking Mandarin), although only partly meeting the criteria above, is still the most widely spoken language in the world.

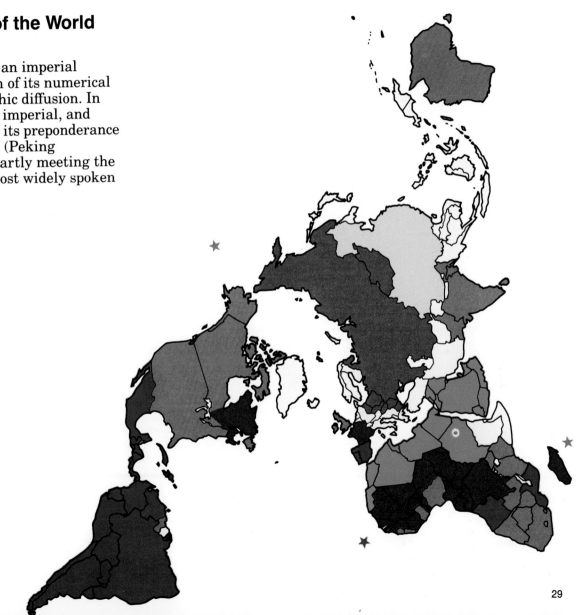

English	
French	
Spanish	
Portuguese	
Arabic	
Russian	
Chinese	

Enduring Traditional Enmities

The traditional enmities noted here are those
that still persist. Franco-German or
Anglo-French rivalry, for example, no longer
has a place here. Conversely, other enmities
based on ancient geohistorical rivalries
continue to revive latent tensions and, often
hidden behind ideological arguments, fuel
more or less open antagonisms.

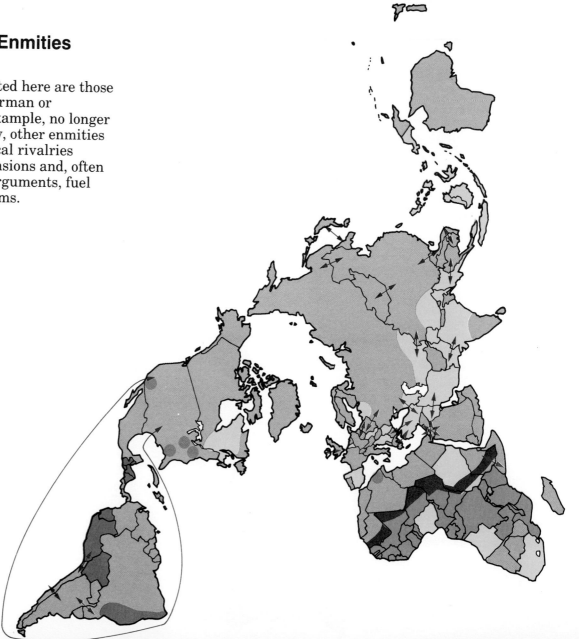

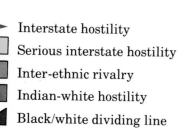

↔ Interstate hostility

Serious interstate hostility

Inter-ethnic rivalry

Indian-white hostility

◣ Black/white dividing line

Traditional Enmities*

ASIA:

China–USSR
China–Vietnam
Vietnam–Khmers
Thailand–Burma
China–Mongols
India–Pakistan
Korea–Japan
Thailand–Cambodia

MIDDLE EAST:

Although of recent origin, the enmity between the Arab countries and Israel has all the characteristics of a lasting rivalry.
Syria–Turkey (fueled by the claim to the *sanjak* of Alexandretta).

AFRICA:

A north–south line through the Sahel divides the black African populations to the south and the Arab and Saharan populations to the north along an ancient cleavage based on slavery. See, for example, Chad, southern Sudan, and so on.

In the Horn of Africa: Ethiopia–Somalia (enmity formerly based on religious rivalry at the present time based on rival nationalisms).

In sub-Saharan Africa, ethnic rivalries are legion. Among those that have been particularly active: Tutsi–Hutu relations in Burundi and Rwanda, or the better-known case of the Ibo of Biafra.

In southern Africa, ethnic and tribal strategies play an important role. Angola: UNITA represents the Ovimbundu, who are opposed to the Luanda government, which rests on the alliance between the Kimbundu and the Bakongo. In Namibia, SWAPO is supported mainly by the Ovambo. In Zimbabwe, there is the rivalry between the Shona and the Ndebele, a source of conflict that could lead to civil war.

In South Africa, Pretoria uses and fosters tribal rivalries, especially between the Zulu and other ethnic groups.

AMERICA:

Brazil–Argentina

Latin America–United States. (The resentment, arising from humiliation, of some groups in Latin America toward the United States is ambiguous to the extent that the interests of privileged strata, however nationalistic they may be, coincide with those of the United States.)

Indians–Hispanics (Peru, Ecuador, Bolivia, Guatemala, and so on).

EUROPE:

Bulgaria–Turkey, Bulgaria–Yugoslavia
Greece–Turkey, Yugoslavia–Albania
USSR–Turkey
Romania–Hungary (claims relating to the Hungarians in Transylvania)
Poland–USSR
Poland–Germany

---

*Interstate enmities not including minority problems. This list is far from complete.

THE HISTORICAL CONTEXT OF THE CONTEMPORARY WORLD

Europeanization of the World at the Beginning of the Twentieth Century

In an uninterrupted process, the expansion of Europe from the sixteenth century onward resulted in the occupation by Europe of the whole of the American continent. During the nineteenth century and until after World War I, European imperialism extended its domination over the whole globe, with the partial exception of Japan. The global superiority of Europe in both technology and *ideas* was at the time total.

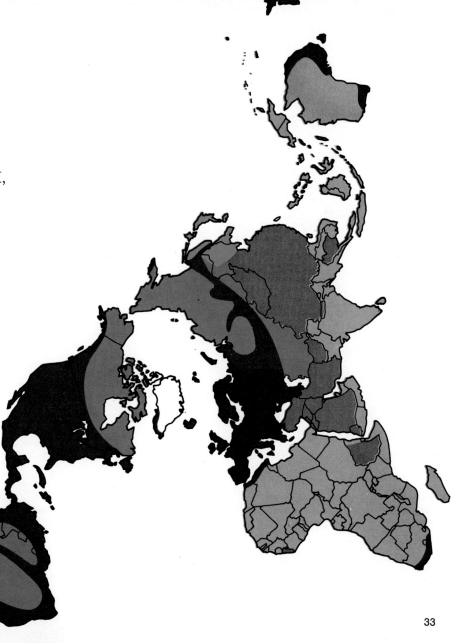

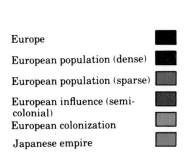

Europe

European population (dense)

European population (sparse)

European influence (semi-colonial)

European colonization

Japanese empire

Territorial Changes in Europe Following the 1914–1918 War

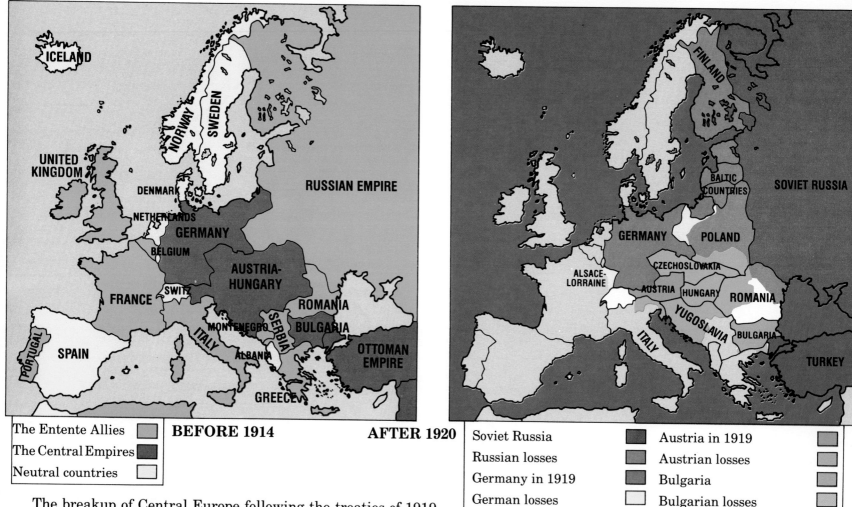

BEFORE 1914

The Entente Allies
The Central Empires
Neutral countries

ICELAND
UNITED KINGDOM
NORWAY
SWEDEN
DENMARK
NETHERLANDS
GERMANY
BELGIUM
FRANCE
SWITZ
ITALY
SPAIN
PORTUGAL
RUSSIAN EMPIRE
AUSTRIA-HUNGARY
MONTENEGRO
SERBIA
ALBANIA
ROMANIA
BULGARIA
OTTOMAN EMPIRE
GREECE

AFTER 1920

Soviet Russia
Russian losses
Germany in 1919
German losses
Austria in 1919
Austrian losses
Bulgaria
Bulgarian losses
Turkey

FINLAND
BALTIC COUNTRIES
SOVIET RUSSIA
GERMANY
POLAND
ALSACE-LORRAINE
CZECHOSLOVAKIA
AUSTRIA
HUNGARY
ROMANIA
ITALY
YUGOSLAVIA
BULGARIA
TURKEY

The breakup of Central Europe following the treaties of 1919–1920 ended traditional German influence in this area and resulted, among other things, in the formation of nation-states based on the ideas of Woodrow Wilson. These states were formed mostly at the expense of the Austro-Hungarian Empire and Russia. By 1945, with few exceptions, all the new states had become part of the Soviet sphere.

The Colonial World between the Two World Wars

Although the states of Latin America were formally fully sovereign, the United States enjoyed almost total economic domination there. Similarly, the few formally independent states in Asia and Africa were often more like semi-colonies, with variations that must not be allowed to obscure their political autonomy.*

*The Union of South Africa became independent in 1910 and obtained a mandate over South-West Africa (Namibia) after World War I.

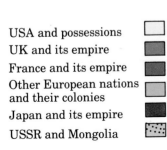

USA and possessions

UK and its empire

France and its empire

Other European nations and their colonies

Japan and its empire

USSR and Mongolia

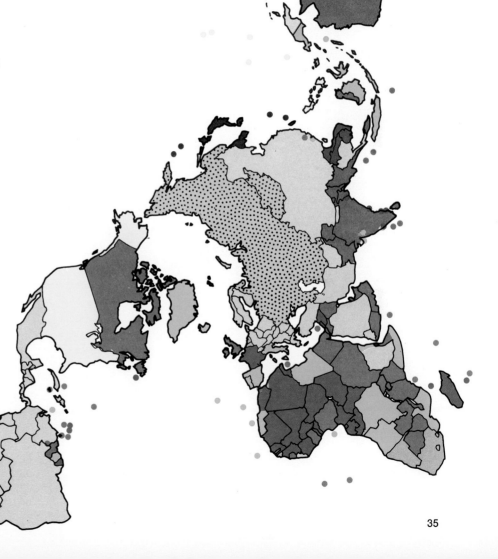

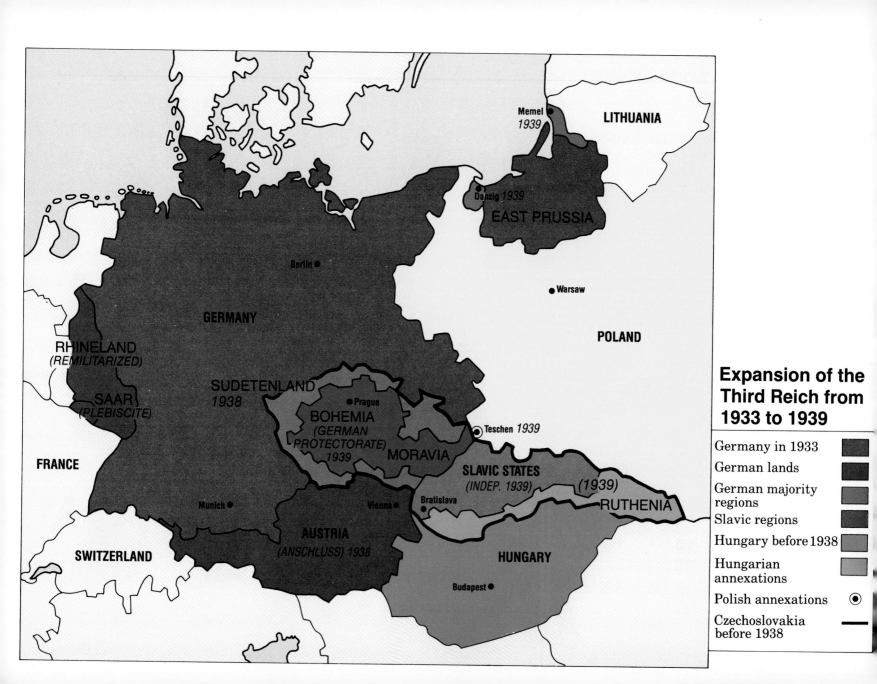

Memel
1939

LITHUANIA

Danzig 1939

EAST PRUSSIA

Berlin ●

● Warsaw

GERMANY

POLAND

RHINELAND
(REMILITARIZED)

SUDETENLAND
1938

SAAR
(PLEBISCITE)

BOHEMIA
(GERMAN
PROTECTORATE)
1939

● Prague

Teschen 1939

FRANCE

MORAVIA

SLAVIC STATES
(INDEP. 1939)

(1939)

Munich ●

Vienna ●

Bratislava
●

RUTHENIA

AUSTRIA
(ANSCHLUSS) 1938

SWITZERLAND

HUNGARY

Budapest ●

**Expansion of the
Third Reich from
1933 to 1939**

Germany in 1933

German lands

German majority
regions

Slavic regions

Hungary before 1938

Hungarian
annexations

Polish annexations ◉

Czechoslovakia
before 1938

Japanese Expansion, 1920–1940

Japan in 1920	■
Expansion 1937-39	■
Protectorate 1932	■

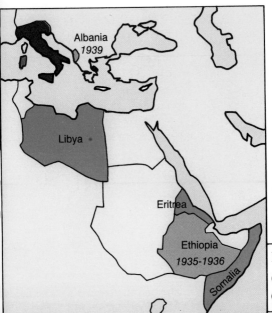

Italy's Colonial Expansion, 1920–1939

Italy	■
Colonies 1920	■
Conquests 1935-39	■

Expansion of the Axis Powers between the Two World Wars

GERMANY

- 1934—reunion of the Saar
- 1936—remilitarization of the Rhineland
- 1938 (March)—annexation of Austria
- 1938 (October)—annexation of the Sudetenland (Munich Conference)
- 1939 (March)—breakup of Czechoslovakia (annexation of Bohemia-Moravia)
- 1939 (March)—annexation of Memel (Lithuania)
- 1939 (September)—invasion of Poland

ITALY

- 1924—annexation of the city of Fiume (Istria)
- 1935–1938—colonization of Cyrenaica (Libya)
- 1935–1936—conquest of Ethiopia
- 1939 (April)—invasion of Albania

JAPAN

- 1931—occupation of Manchuria
- 1935—annexation of Chahar and Suiyuan (Chinese Mongolia)
- 1937–1939—occupation of northeast China, the lower Yangtse valley, and the southern coastal areas

The World at War—1942

This map shows the territorial imbalance
between the Axis Powers at the time of their
greatest expansion (late 1942) and the Allies,
their possessions, and the states sympathetic
to them.

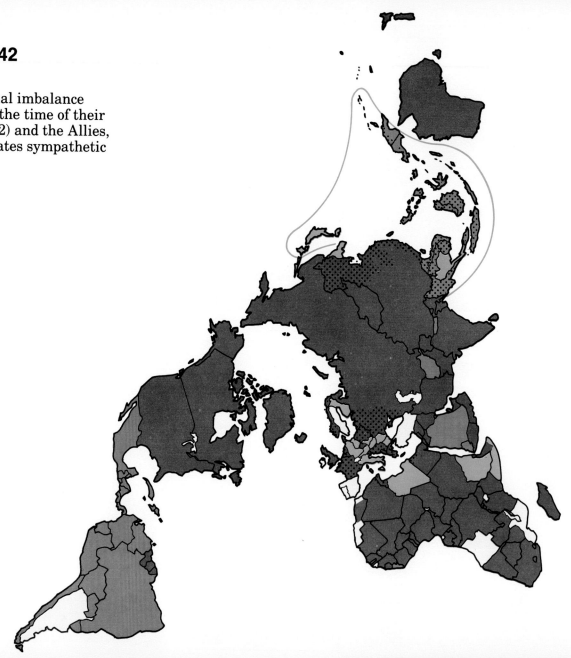

Allied nations

Favorable to Allies

Axis nations

Neutral

Maximum expansion
of Axis nations

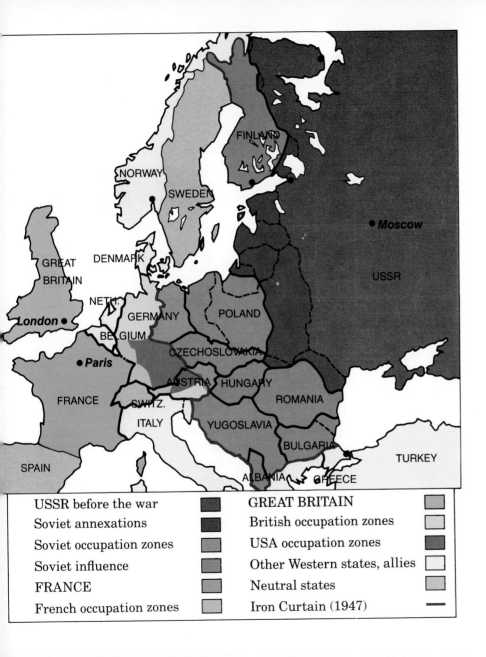

Europe after World War II

The USSR gained some 600,000 square kilometers in the west at the expense of Poland, the Baltic countries, Romania, Czechoslovakia, and Finland. In terms of square kilometers, it recovered exactly what Russia had lost by the Treaty of Brest-Litovsk (1918).

The great loser was Germany, divided into occupied zones and then into two states. The Cold War began in 1948 and froze this division. The Helsinki Accords (1976) formally confirmed Soviet domination in Eastern Europe.

Map labels:

NORWAY, SWEDEN, FINLAND, • Moscow, DENMARK, USSR, GREAT BRITAIN, NETH., GERMANY, POLAND, • London, BELGIUM, CZECHOSLOVAKIA, • Paris, AUSTRIA, HUNGARY, FRANCE, SWITZ., ROMANIA, ITALY, YUGOSLAVIA, BULGARIA, TURKEY, SPAIN, ALBANIA, GREECE

Legend:

- USSR before the war
- Soviet annexations
- Soviet occupation zones
- Soviet influence
- FRANCE
- French occupation zones
- GREAT BRITAIN
- British occupation zones
- USA occupation zones
- Other Western states, allies
- Neutral states
- Iron Curtain (1947) —

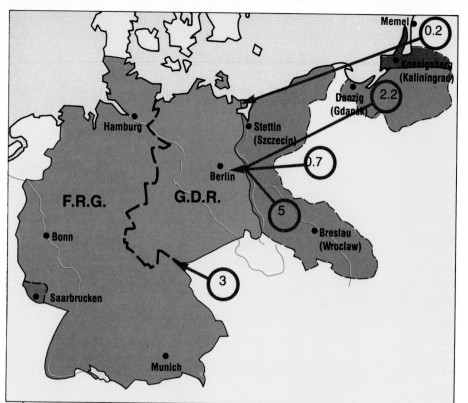

Germany 1945

Germany, which had an area of 540,000 sq. km. before 1914 and 474,000 sq. km. before 1938, was reduced to 248,000 sq. km. for the Federal German Republic and 108,000 sq. km. for the German Democratic Republic. Some 12 million Germans were expelled from various states in Central Europe and returned to occupied Germany. Between 1947 and 1961, some 4 million East Germans sought refuge in West Germany. For fifteen years the *Ostpolitik,* which accepts the fact of the division, has reflected the particular attitude of West Germany and of a large section of its public opinion toward problems of war and peace.

Poland

Before 1939: 388,000 sq. km.
Today: 312,000 sq. km.
Six million Poles were deported and exterminated, including 3 million Jews. One million Poles were expelled from the eastern regions. The Soviet Union's territorial ambitions in the west led the Allies at Yalta (February 1945), and especially at Potsdam (July–August 1945), to compensate Poland for Russian annexation of the eastern Polish provinces by granting Poland Silesia and part of Pomerania and East Prussia.

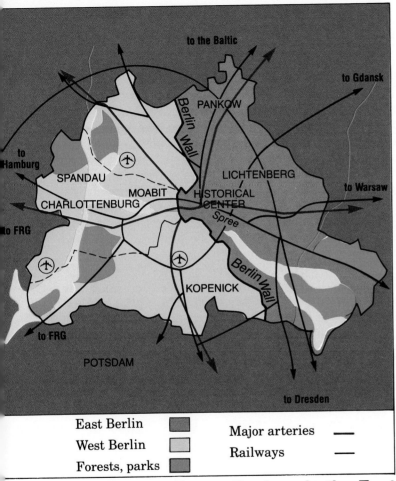

East Berlin	▨
West Berlin	▨
Forests, parks	▨

Major arteries	—
Railways	—

West Berlin, an Enclave in the East

West Berlin: An Anomaly

West Berlin, situated in the middle of the German Democratic Republic, 110 km. from West Germany, was for a long time (1947–1961) an abscess in the body of the Cold War. After the building of the wall by the GDR, the status quo gradually became stabilized.

CHRONOLOGY:

1949—blockade of Berlin by the USSR; American airlift
1961—construction of the wall cutting off West Berlin and preventing the exodus of East Germans

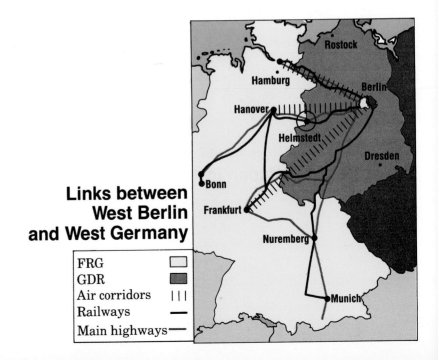

Links between West Berlin and West Germany

FRG	▢
GDR	▨
Air corridors	|||
Railways	—
Main highways	—

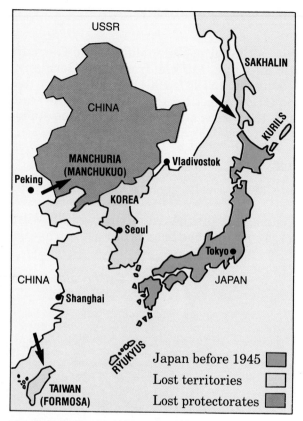

Defeated Japan in 1945

USSR

SAKHALIN

CHINA

KURILS

MANCHURIA
(MANCHUKUO)

Peking

Vladivostok

KOREA

Seoul

Tokyo

CHINA

JAPAN

Shanghai

RYUKYUS

Japan before 1945

Lost territories

Lost protectorates

TAIWAN
(FORMOSA)

Japan

The Japanese empire (630,000 sq. km., including 380,000 sq. km. for Japan proper) collapsed after the atomic bomb attacks on Hiroshima and Nagasaki in 1945. The USSR, which had declared war on Japan at the last moment, annexed strategic positions in the north of the archipelago in the Kuril Islands and the southern half of Sakhalin. China recovered its sovereignty over Manchuria and the island of Taiwan.

Japan was demilitarized and underwent a process of institutional democratization during the American occupation. In Micronesia, the United States secured trusteeship over the formerly Japanese island chains of the Marianas, the Carolines, and the Marshalls. In the late 1970s the United States handed back the Ryukyu archipelago (Okinawa). The Greater East Asia Co-Prosperity Sphere, envisaged militarily in the 1930s, appears to be becoming a reality today economically, although the impact of Japanese commercial penetration is worldwide.

The West's Perception, at the Beginning of the Cold War (1948–1952), of Soviet and Communist Expansionism Worldwide

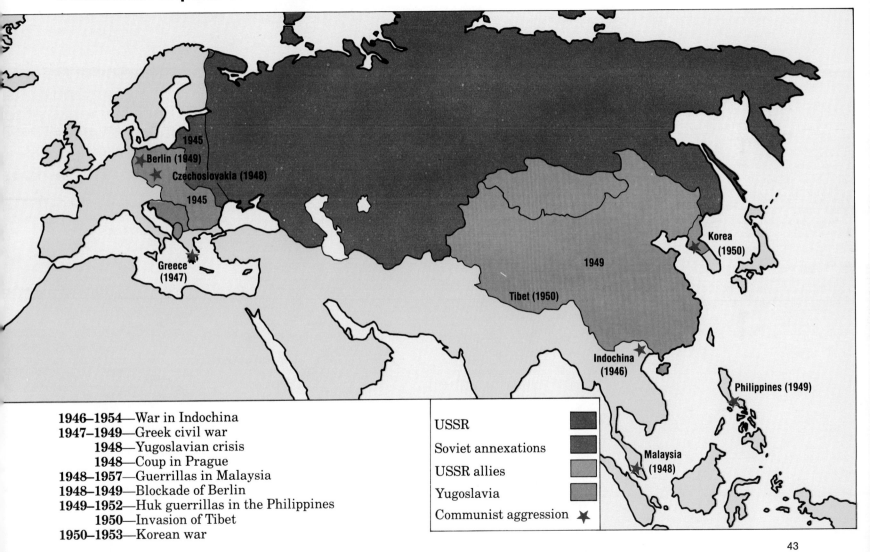

1945
Berlin (1949)
Czechoslovakia (1948)
1945
Greece (1947)
1949
Tibet (1950)
Korea (1950)
Indochina (1946)
Philippines (1949)
Malaysia (1948)

1946–1954—War in Indochina
1947–1949—Greek civil war
     1948—Yugoslavian crisis
     1948—Coup in Prague
1948–1957—Guerrillas in Malaysia
1948–1949—Blockade of Berlin
1949–1952—Huk guerrillas in the Philippines
     1950—Invasion of Tibet
1950–1953—Korean war

USSR
Soviet annexations
USSR allies
Yugoslavia
Communist aggression ★

43

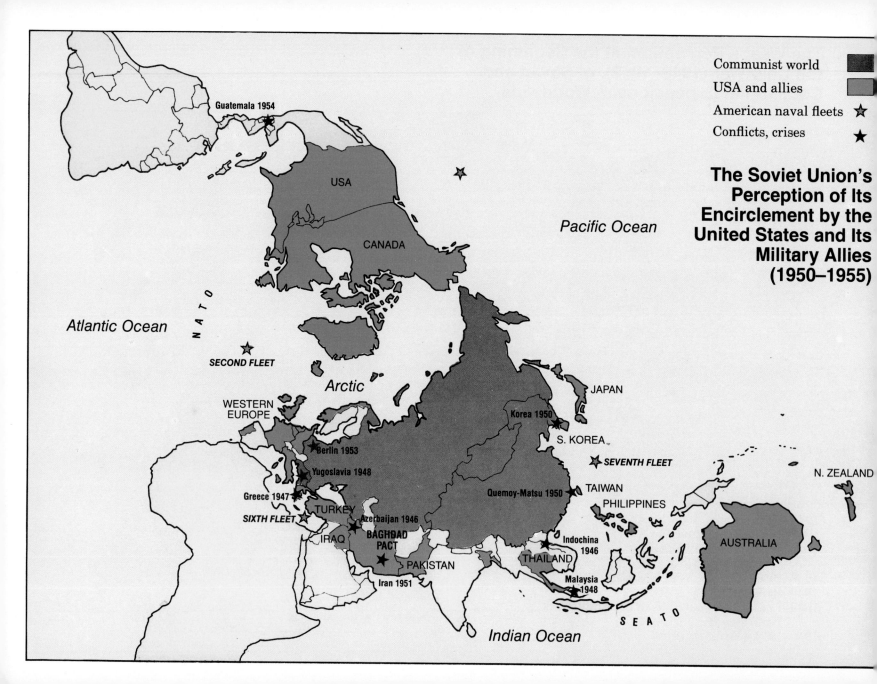

Communist world
USA and allies
American naval fleets ☆
Conflicts, crises ★

The Soviet Union's Perception of Its Encirclement by the United States and Its Military Allies (1950–1955)

Guatemala 1954

USA

CANADA

Pacific Ocean

Atlantic Ocean

SECOND FLEET

Arctic

WESTERN EUROPE

JAPAN

Korea 1950

S. KOREA

Berlin 1953

SEVENTH FLEET

Yugoslavia 1948

Greece 1947

Quemoy-Matsu 1950

TAIWAN

PHILIPPINES

N. ZEALAND

TURKEY

Azerbaijan 1946

SIXTH FLEET

IRAQ

BAGHDAD PACT

Indochina 1946

AUSTRALIA

THAILAND

PAKISTAN

Iran 1951

Malaysia 1948

S E A T O

Indian Ocean

Decolonization and the New States, 1945–1983

From the end of World War II, first in Asia and then a decade later in Africa, in a process that was at times violent, national liberation movements brought about the independence of over a hundred new states. Namibia (formerly South-West Africa) remains at present under South African rule.

Independent 1945

Independent since 1945

Dependent in 1983

Guyana (France)

Namibia (South Africa)

Chronology of Decolonizations in Southeast Asia and in Africa

THE NEAR EAST

	Date of Independence
Yemen	1918
Saudi Arabia	1926
Iraq	1932
Jordan (Trans-Jordan)	1946
Libya, Syria	1946
Israel (partition of Palestine)	1948

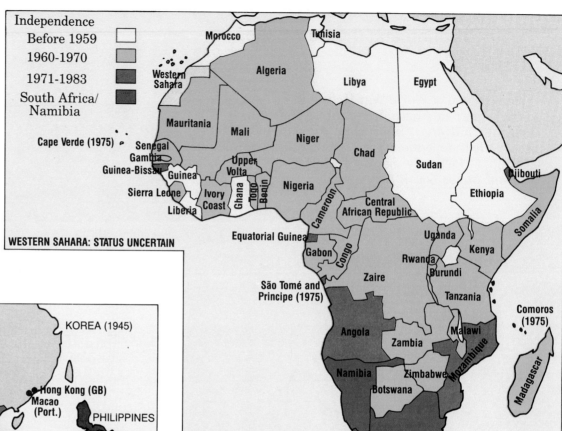

Independence
Before 1959 (white)
1960-1970 (light gray)
1971-1983 (medium gray)
South Africa/Namibia (dark)

Cape Verde (1975)

WESTERN SAHARA: STATUS UNCERTAIN

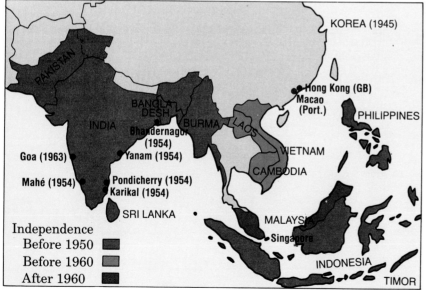

KOREA (1945)

Hong Kong (GB)
Macao (Port.)

Independence
Before 1950
Before 1960
After 1960

Cyprus	1960
Kuwait	1961
South Yemen	1967
Oman, Bahrain, the Emirates	1971

Several archipelagos and islands—and Guyana—in the Caribbean are not independent.

Conflicts in the World since 1945

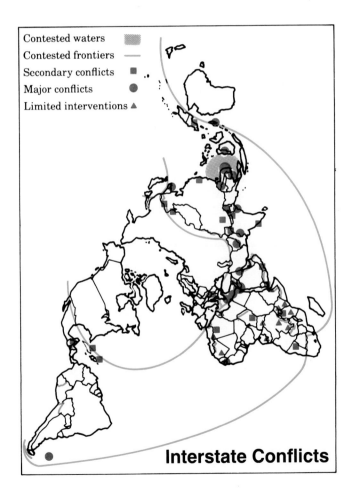

Contested waters

Contested frontiers

Secondary conflicts ■

Major conflicts ●

Limited interventions ▲

Interstate Conflicts

There have been over a hundred significant conflicts during the period 1945–1983.

To classify them, the following typology may be proposed:

- Interstate conflicts (conventional wars)
- Liberation wars (in a colonial context or one of foreign occupation)
- Internal conflicts (civil wars arising out of class, ethnic, and/or religious conflicts)

It is estimated that these conflicts have generated, in the period 1945–1982, some 13.5 million victims, an annual rate of about 350,000. (We do not deal here with the wave of terrorist attacks, whether trans-state or not, which are often a substitute for guerrilla warfare but cause a very limited number of victims. When this form of conflict is used as the sole means, it is more than anything else a matter of psychological warfare.)

Ten of these wars, not necessarily those that have had most impact on public opinion, have alone involved over 10 million victims, three-quarters of the total. These are the two Indochina wars (1946–1975), the Indo-Pakistan wars (1947–1949 and 1971 Bangladesh), the Korean war, the Algerian war, the civil war in Sudan, the massacres in Indonesia (1965), and the Biafran war.*

Many conflicts have resulted in significant numbers of refugees: Palestinians, subcontinental Indians, Indochinese, Ethiopians, Afghanis, Central Americans, and so on.

---

*Apart from wars and guerrilla wars, other types of political conflict have resulted in large numbers of victims. For example in Cambodia (1975–1978), in China during the Cultural Revolution, in Equatorial Guinea under Macias or in Uganda under Idi Amin Dada, in Argentina, in Rwanda (1960–1965), in Burundi (1972–1973), and so on.

I. INTERSTATE CONFLICTS

A. MAJOR CONFLICTS:
India-Pakistan (1947–1949)
Arab-Israeli (1948–1949)
Korea (1950–1953)
Israel-Egypt (1956)
India-Pakistan (1965)
Vietnam (1965–1973)—massive intervention by the USA*
Arab-Israeli (1967)
India-Pakistan (1971)—Bangladesh
Arab-Israeli (1973)
Vietnam-China (1979)
Iran-Iraq (1980–   )
Great Britain-Argentina (1982)—Falkland Islands

B. SECONDARY CONFLICTS:
China-Taiwan (1950)—Quemoy-Matsu
China-Tibet (1950–1951)
Guatemala-Honduras (1954)—Operation CIA
India-China (1959)—Ladakh
Netherlands-Indonesia (1960–1962)—New Guinea
India-China (1962)—Assam
Indonesia-Malaysia (1963)—Sarawak, Borneo
Algeria-Morocco (1963)
China-USSR (1969)—Ussuri
Salvador-Honduras (1969)
Greece-Turkey (1974)—Cyprus
Syria-Lebanon (1976)—occupation
Indonesia-East Timor (1976)—annexation by Indonesia

Somalia-Ethiopia (1977–1978)
Vietnam-Cambodia (1978)
North Yemen-South Yemen (1979)
Israel-Lebanon (1982)—PLO

C. INTERVENTIONS:
Suez, French-British intervention (1956)
Budapest, Soviet intervention (1956)
Lebanon, U.S. intervention (1958)
Mauritania, French intervention (1961)
Cuba (1961)—Playa Girón
Bizerte, French intervention (1961)
Goa (Portuguese possession), Indian intervention (1961)
Zaire, Belgian intervention (1961 and 1964)
Gabon, French intervention (1964)
Uganda-Kenya-Tanzania, British intervention (1964)
Santo Domingo, U.S. intervention (1965)
Prague, Soviet intervention (1968)
Cambodia, U.S. intervention (1970)
Jordan, royal forces against the PLO (1970)
Angola, interventions by South Africa, Zaire, and especially Cuba (1975–1976)
Shaba (Zaire), French and Moroccan interventions (1977)
Djibouti, French intervention (1976–1977)
Ethiopia, Cuban intervention (1977)
Kolwezi (Zaire), French intervention (1978)
Chad, numerous interventions by French forces (1968–1980)
Uganda, Tanzanian intervention (1979)
Angola, South African interventions (1980, 1981, 1982)
Central Africa, French intervention (1979)
Chad, Libyan intervention (1980)

---

*Air strikes in North Vietnam; civil war in South Vietnam.

Gambia, Senegalese intervention (1980)
Chad, French intervention (1983)
Grenada, American intervention (1983)

II. LIBERATION MOVEMENTS FOR INDEPENDENCE DIRECTED AGAINST FOREIGN DOMINATION OR OCCUPATION

Palestine (Zionist movement), against Great Britain (1945–1947)
Indochina war (Vietnam), against France (1946–1954)
Laos (Pathet-Lao), against France (1946–1954)
Indonesia, against the Netherlands (1946–1949)
Malaysia, against Great Britain (1948–1957)
Kenya (Mau Mau insurrection), against Great Britain (1952–1954)
Tunisia, against France (1952–1956)
Morocco, against France (1953–1956)
Algeria, against France (1954–1962)
Cyprus, against Great Britain (1955–1959)
Cameroon, against France (1957–1960)
Belgian Congo (1958–1960)
Angola, against Portugal (1961–1974)
South Yemen, against Great Britain (1963–1967)
Guinea-Bissau, against Portugal (1963–1974)
Palestinians, against Israel, especially since 1967 (1965–   )
Mozambique, against Portugal (1964–1974)
Namibia, against South Africa (1970–   )
Rhodesia/Zimbabwe, against white Rhodesian domination (1972–1979)
East Timor, against Indonesia (1974–   )
Western Sahara, against Morocco (1975–   )

Cambodia, against a regime put in place by Vietnam (1979–   )
Afghanistan, against the Soviet occupation (1979–   )

III. CONFLICTS OVER SECESSION* OR TO GAIN AUTONOMY WITHIN ESTABLISHED STATES

Azerbaijan and the Kurdish republic of Mahabad, Iran (1946)
Burma (Karens, etc.) (1948–1954)
Hyderábad, resistance to incorporation into India (1948)
South Moluccas (1950–1952)
Tibet, against China (1955–1959)
Katanga (Zaïre) (1960–1964)
Kurds, in Iraq (1961–1970, 1974–1975, 1979–   )
Eritrea, Ethiopia (1961–   )
South Sudan (1966–1972, 1982–   )
Biafra, Nigeria (1967–1970)
India (Nagas) (1967–1970)
Baluchistan (Pakistan) (1973–1977)
Ogaden (Ethiopia) (1974–   )
Basques in Spain (1975–1981)
Philippine Muslims (1977–   )
Kurds in Iran (1978–   )

---

*As in civil wars (IV), the date of the beginning of the operations is often uncertain.

IV. CIVIL WARS FOUGHT TO CHANGE REGIMES*

China (1945–1949)
Greece (1947–1949)
Huks (Philippines) (1949–1952)
Colombia, chronic upheaval (1953)
Cuba (1956–1959)
South Vietnam (1957–1964; 1973–1975)
Sumatra, insurrection against the centrist regime of
    Jakarta (1957–1958)
Zaire (1960–1965)
Malaysia, (sporadic)
Laos (1960–1975)
Thailand (sporadic)
Cameroon (1960–1966)
Guatemala (1961–1968; 1980–   )
Venezuela (1962–1967)
Yemen, with Egyptian intervention (1962–1967)
Rwanda (1963–1964)
Cyprus, intervention by the U.N. (1963–1964)
Cambodia (1965–1975)
Indonesia (1965)
Uruguay (1965–1973)
Peru (1965, and 1982–   )
Bolivia (1967)
Brazil (1967–1970)
Northern Ireland (Catholics) (1968–   )
Chad (1968–1982)
Dhofar, in Oman, with intervention by Great Britain,
    Iran, and Jordan (1968–1976)
Nicaragua, (1972–1979)
Burundi (1972)
Chile, military repression (1973)
Argentina (1973–1977)
Lebanon, chronic upheaval (1975–1977)
Angola (UNITA), with the aid of South Africa
    (1976–   )

El Salvador (1976–   )
Iran (1978–1979)
Afghanistan (1978–1979)
Mozambique (FMN), with the aid of South
    Africa (1980–   )
Philippines (1980–   )

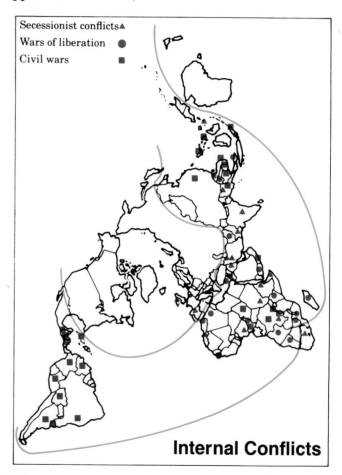

Secessionist conflicts ▲
Wars of liberation ●
Civil wars ■

Internal Conflicts

*For all intents and purposes, all such conflicts since 1945 have taken place in the Third World.

A WORLD OF OCEANS

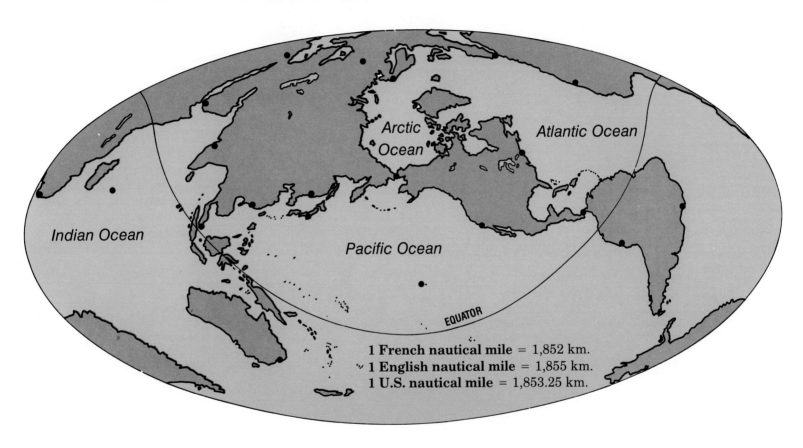

Arctic Ocean

Atlantic Ocean

Indian Ocean

Pacific Ocean

EQUATOR

1 French nautical mile = 1,852 km.
1 English nautical mile = 1,855 km.
1 U.S. nautical mile = 1,853.25 km.

Total surface of the world—510 million sq. km.
Land surface—149 million sq. km. (29%)
Covered by water—361 million sq. km. (71%)

Ocean surfaces
Pacific Ocean—161.7 million sq. km.
Atlantic Ocean—81.6 million sq. km.
Indian Ocean—73.4 million sq. km.
Arctic Ocean—14.3 million sq. km.

Military Dispositions around the Arctic

The Arctic separates the USSR and the United States. Control of the surface of this ocean and its air space is of vital importance. Beneath the frozen stretches of sea there has developed, with nuclear submarines, a subglacial theater of operations of equally vital strategic importance.

Map labels:

Pacific Ocean

KISKA
ADAK SHEMYA
UNALASKA
Aleutians
PETROPAVLOVSK
SOV GAVAN
JAPAN
PRIBILOV
UM KAMTCHASK
KAMCHATKA
NIKOLAYEVSK
VLADIVOSTOK
5000 km
KODIAK
MAGADAN
OKHOTSK
5000 km
ANCHORAGE NOME
ANADYR
VICTORIA
PROVIDENIYA
USA
KOTZEBUE
AMBARTCHIK
YAKOUTSK
VANCOUVER
FAIRBANKS
PEVEK
MAC CHORD
BARROW
6000 km
MALMSTROM
CANADA
Beaufort Sea
EDMONTON
Laptev Sea
TIKSI
MINOT
SAWYER
6500 km
NORDVIK
USSR
G. FORKS
NORILSK
WINNIPEG
CHURCHILL
NORTH POLE
DIKSON
Hudson Bay
NOVOSIBIRSK
THULE
4500 km
Barents Sea
6500 km
4500 km
Novaya Zemlya
VORKOUTA
Frobisher Bay
Baffin Bay
GREENLAND
QUEBEC
Svalbards
SVERDLOVSK
6500 m
LORING
SONDRE
MURMANSK
Kola
ARKHANGELSK
HALIFAX
ARGENTIA
Strait of Denmark
Sea of Norway
NARVIK
KEYFLAVIK
ICELAND
Faeroes
LENINGRAD
Atlantic Ocean
NORWAY
MOSCOW
OSLO
STOCKHOLM
GREAT BRITAIN
HOLY LOCH

Legend:

U.S. and NATO bases	✳
Soviet bases	★
Distances	←
Claims of sovereignty	– – –
Contested area (USSR-Norway)	▨
Svalbard area	⬚
Permanent ice cap	☐
Seasonal ice	☐
Open waters	■

The Arctic: Communications

The Arctic is not well-suited to navigation. With the ports of Murmansk and Archangel, the USSR has a relatively favorable position there.

The Faeroe Islands and the Strait of Denmark, outlets to the Atlantic, are easy to watch. Arctic air space is regularly used by several international airlines.

Legend:

Air routes	
Claims of sovereignty	---
Contested area	
Svalbard Is. area	
Permanent ice cap	
Seasonal ice	
Open waters	

JAPAN

Pacific Ocean

ALEUTIANS

UM KAMTCHASK

PETROPAVLOVSK

VLADIVOSTOK

PRIBILOV

KAMTCHATKA

Sea of Okhotsk

NIKOLAYEVSK

NEW YORK-JAPAN

ANADYR

MAGADAN

ANCHORAGE

Bering Strait

AMBARTCHIK

VANCOUVER

FAIRBANKS

ALASKA

Convention of 1867

PEVEK

YAKOUTSK

USA

CANADA

EDMONTON

Beaufort Sea

TIKSI

Lena

COPENHAGEN-JAPAN

LONDON-VANCOUVER

MOSCOW-JAPAN

USSR

VICTORIA

NORDVIK

LOS ANGELES-COPENHAGEN

North Pole

WINNIPEG

CHURCHILL

ELLESMERE

NORILSK

Ienissei

Hudson Bay

DIKSON

Barents Sea

NOVOSIBIRSK

Frobisher Bay

GREENLAND

SVALBARDS

VORKOUTA

Ob

QUEBEC

Baffin Bay

Accord of 1973

Treaty of 1920

SVERDLOVSK

HALIFAX

GODTHAB

MURMANSK

KOLA

ARKHANGELSK

Denmark Strait

NARVIK

REYKJAVIK

ICELAND

FAEROES

LENINGRAD

Atlantic Ocean

NORWAY

STOCKHOLM

MOSCOW

OSLO

GREAT BRITAIN

GLASGOW

The Bering Strait

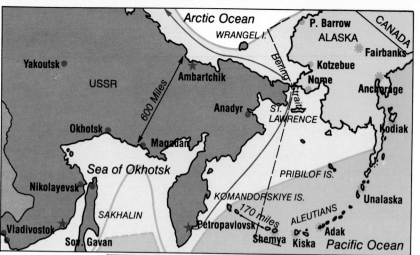

Permanent ice cap
Icebound (6-9 months)
Open water
Iceberg zones
USA and allies
USSR and allies
Soviet radar system
Arctic sea lane
NATO radar system

U.S. and Allied bases
U.S. support points
Soviet bases
Soviet support points
Iron ore
Other mineral ores
Airfields
Distances in nautical miles

North Cape Passage

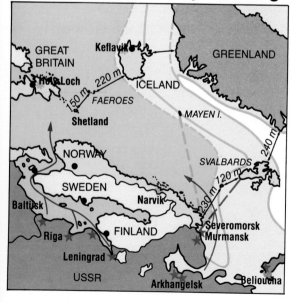

Kola Peninsula

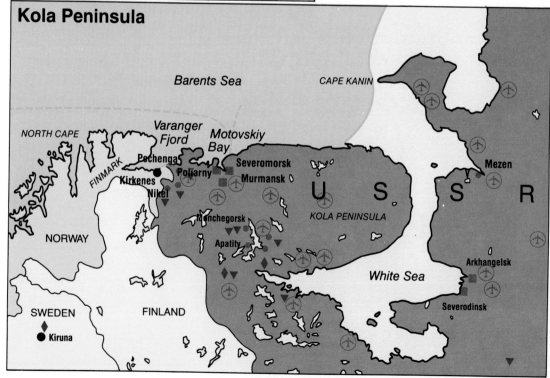

Outlets from the Arctic

Since the early 1920s, the coastal states have pushed the limits of their territorial waters well beyond those then accepted. There is a dispute between the USSR and Norway about this covering an area of 150,000 sq. km. around the Svalbard Islands. By the Treaty of Paris (1920), 41 signatory states (including the USSR) share equal rights to mine coal in the Svalbard Islands—which, however, remain under Norwegian authority.

TOWARD THE ATLANTIC

This access route is essential to the northern Soviet fleet based in the region of Murmansk and the White Sea. The polar basin between Greenland and the Svalbard Islands (Spitsbergen) is icebound most of the time and is difficult to navigate. The Barents Sea between North Cape and the Svalbard Islands is the only passage in open waters. Norway's position at North Cape is strategically vital, but highly vulnerable.

TOWARD THE PACIFIC

The Bering Strait (40 miles) is also icebound for more than six months a year. Its narrowness and shallowness (about 40 m.) make it easy to watch and blockade. Farther south, the string of the Aleutian (USA) and the Komandorskie islands (USSR) complete control of this outlet from the Arctic toward the Pacific.

THE NORTHERN SEA ROUTE

This route is open two or three months a year and makes it easier for the USSR to develop northern Siberia, and provides a shorter sea route between Murmansk and Vladivostok. There is no similar route open along the northern coasts of Canada. Only seasonal navigation is possible in Hudson Bay and the Baffin Sea.

Alaska

Alaska, the poleward extension of the United States, (1.5 million sq. km. and 400,000 inhabitants) is of major strategic significance and contains large reserves of oil and natural gas. Anchorage is heavily used by transpolar airlines.

Outlets

Barents Channel:
    North Cape → 280 NM ← Ours I. → 120 NM → Svalbard Is.

Denmark Strait: 150 NM
    Greenland → Svalbard: 240 NM
    Iceland → Faeroe Is.: 200 NM
    Faeroe Is. → Shetland Is.: 150 NM

Distances
    Bering Strait ↔ Denmark Strait: 3,000 NM
    Bering ↔ Murmansk (northern sea route): 3,800 NM

NM: Nautical miles.

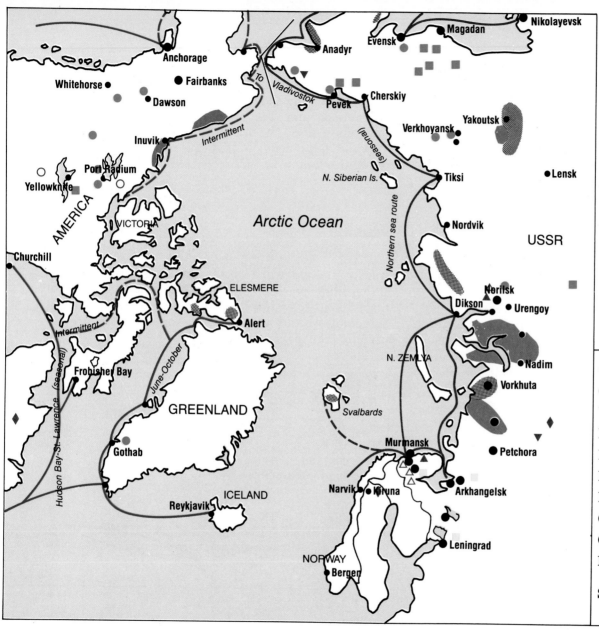

The Arctic: Enormous Resources, Difficult to Exploit

Nikolayevsk
Magadan
Evensk
Anadyr
Anchorage
Whitehorse
Fairbanks
Dawson
To Vladivostok
Cherskiy
Pevek
Yakoutsk
Verkhoyansk
Inuvik
Intermittent
Port Radium
Yellowknife
N. Siberian Is.
Lensk
AMERICA
VICTORIA
Tiksi
Arctic Ocean
Northern sea route (seasonal)
Nordvik
USSR
Churchill
ELESMERE
Norilsk
Dikson
Urengoy
Alert
Intermittent
N. ZEMLYA
Nadim
Frobisher Bay
June-October
Vorkhuta
Hudson Bay–St. Lawrence (Seasonal)
GREENLAND
Svalbards
Gothab
Murmansk
Petchora
Narvik
Kiruna
Arkhangelsk
Reykjavik
ICELAND
Leningrad
NORWAY
Bergen

Oil and natural gas	▨
Coal	▨
Uranium	○
Iron	◆
Iron alloys	▼
Nickel	▲
Bauxite	▨
Gold	■
Copper	△
Nonferrous metals	●
Sea routes	Soviet
	U.S.

The Atlantic: A Western Ocean

The North Atlantic is the "inner sea" of the Euro-American world and the countries in its geopolitical orbit.

In the absence of any coastal maritime power, the South Atlantic remains an area essentially controlled by Europe and the United States.

Except for Iceland and the Cape Verde islands, the islands in the sea are all still under the control of Great Britain and other NATO countries.

The Canaries, Madeira, and especially the Azores are valuable staging points in the North Atlantic.

The USSR, whose fleet has grown considerably, was for a long time not present on this ocean, but today it has access to facilities, especially in Cuba and Angola.

Map labels

GREENLAND
ICELAND
CANADA
UNITED KINGDOM
3570 nm
FRG
FRANCE
UNITED STATES
NEW YORK
3150 nm
PORTUGAL SPAIN
Azores (Port.)
Str. of Gibraltar
Mediterranean
Bermuda (UK)
Madeira (Port.)
6800 nm
MEXICO
2000 nm
Canary Is. (Sp.)
CUBA
Puerto Rico (US)
Caribbean
Antilles
Cape Verde Is.
DAKAR
NICARAGUA
NIGERIA
VENEZUELA
PANAMA
4800 nm
1750 nm
EQUAT. GUINEA
S. TOMÉ
F. De Noronha (BR)
BRAZIL
★ Ascension (UK)
St. Helena (UK)
ANGOLA
6200 nm
3300 nm
Pacific Ocean
SOUTH AFRICA
THE CAPE
ARGENTINA
2400 nm
Tristan da Cunha (UK)
Gough (UK)
Falklands (UK)
South Georgia (UK)

Legend

- Pro-Soviet states
- USA and NATO members
- Other Western allies
- European neutrals
- African states
- Latin American states
- Frozen in winter
- 200-mile limit
- U.S. and NATO bases ★
- Soviet bases ✳
- Distances

The Atlantic Ocean: An Economic Crossroads

The main industrial areas of North America and Western Europe remain heavily dependent on supplies of raw materials that come via the Atlantic. The security of sea routes and the various key access points is thus vital for the Western powers.

In terms of the tonnage transported and the number of ships using it, the Atlantic is the busiest ocean. Similarly, North Atlantic air space has the heaviest commercial aviation traffic.

The Busiest Ocean

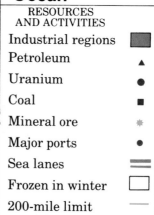

RESOURCES AND ACTIVITIES

Industrial regions	▇
Petroleum	▲
Uranium	●
Coal	■
Mineral ore	✳
Major ports	●
Sea lanes	═══
Frozen in winter	▢
200-mile limit	──

Hudson Bay

Seasonal route

Oslo

Rotterdam

Montreal

London Hamburg

NORTHEAST GREAT LAKES

NORTHWEST EUROPE

Bilbao

Boston
New York

Genoa

AZORES

Lisbon

Gibraltar

BERMUDA

MADEIRA

Casablanca

New Orleans

CANARY IS.

Western USA

Colon

PUERTO RICO

ANTILLES

CAPE VERDE

Dakar

Panama Canal

Maracaibo

Lagos

Abidjan

EQUAT. GUINEA
S. TOMÉ

Peru

FERNANDO DE NORONHA

ASCENSION

Luanda

Salvador

ST. HELENA

BRAZILIAN TRIANGLE

Rio
Santos

Buenos Aires

TR. DA CUNHA

Capetown

GOUGH

Asia-Near East

Chile

FALKLANDS

Str. of Magellan

S. GEORGIA

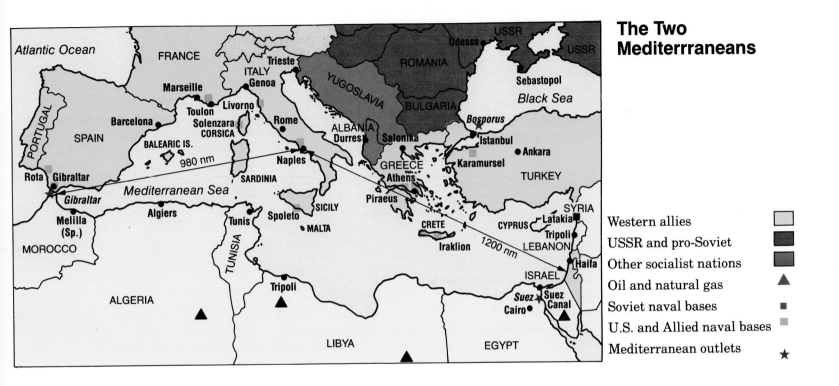

The Two Mediterraneans

Atlantic Ocean

FRANCE
ITALY
Trieste
Genoa
Marseille
Toulon
Livorno
Solenzara
CORSICA
Barcelona
Rome
SPAIN
BALEARIC IS.
Naples
980 nm
SARDINIA
Rota Gibraltar
Gibraltar
Mediterranean Sea
Melilla
(Sp.)
Algiers
SICILY
Spoleto
MOROCCO
Tunis
MALTA
TUNISIA
Tripoli
Iraklion
ALGERIA
Tripoli
LIBYA

PORTUGAL

YUGOSLAVIA
ALBANIA
Durres
Salonika
GREECE
Athens
Piraeus
CRETE
1200 nm

USSR
Odessa
ROMANIA
BULGARIA
Black Sea
Sebastopol
Bosporus
Istanbul
Karamursel
Ankara
TURKEY
SYRIA
Latakia
CYPRUS
Tripoli
LEBANON
Haifa
ISRAEL
Suez
Cairo
Suez Canal
EGYPT

USSR

Legend:
- Western allies
- USSR and pro-Soviet
- Other socialist nations
- Oil and natural gas ▲
- Soviet naval bases ■
- U.S. and Allied naval bases ■
- Mediterranean outlets ★

THE MEDITERRANEAN

Since 1945, the strategic importance the Mediterranean had during the period of European hegemony has rapidly declined. Decolonization and the Arab–Israeli conflicts, which led on several occasions to the closure of the Suez Canal, have considerably reduced the level of its commercial activity. The main route for petroleum products today is around South Africa. Militarily, long-range strategic weapons have transformed this easily blockaded sea into a trap for fleets.

The role of the Mediterranean has become regional, even though at present the conflict in the Middle East and the upkeep of NATO's military forces in Italy, Greece, and Turkey require the presence of naval forces. Since its disengagement from Egypt, the USSR now has only one staging area—in Syria.

THE CARIBBEAN BASIN

The Caribbean forms an "American Mediterranean." Traditionally, the United States has perceived this area as strategically vital, and it has numerous bases there, the main ones being at Panama, in the Canal Zone, in Puerto Rico, and at Guantanamo (Cuba). In the past, and recently, the United States has frequently intervened militarily in Central America and the Caribbean.

Compared to North and South America, the proliferation of ministates and the large number of small islands (a minority of which are not independent) make this area particularly vulnerable. In the last twenty years, more than a dozen new states have won independence. Since it was built, the Panama Canal has been primarily a passageway between the east and west coasts of the United States.

The Caribbean basin, dominated by the United States (except for Cuba, which appears as a troublemaker), is important for its oil and as a communications nexus. It has recently seen the emergence of two regional powers, Venezuela and Mexico. The states in this basin have been benefiting from favorable economic conditions assured by the United States, as well as from the low oil prices granted them by Venezuela and Mexico. Miami has become the real center of the Caribbean basin.

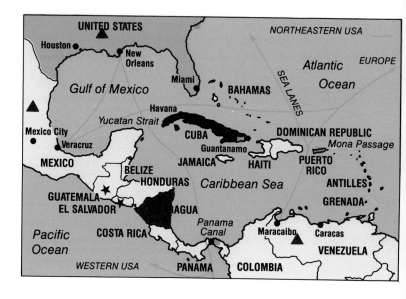

USA	
USSR and pro-Soviet	
Oil and natural gas	▲
Soviet bases	▪
U.S. bases, military installations	
Guerrillas	★

The Indian Ocean

A MAJOR AREA OF CONFLICT

Because of the flow of petroleum from the Persian Gulf and the instability or fragility of many of the coastal states, the Indian Ocean is potentially a major area of conflict.

From its important base on Diego Garcia, the United States is endeavoring, with its allies (Great Britain, France, etc.), to organize an operational strike force in this sensitive area. In addition to the 32 American ships, there are some 30 British vessels. France has 7 warships at Djibouti.

Since 1970, the deployment of Soviet forces, especially naval forces, has increased considerably. In Ethiopia and in South Yemen, and on Socotra, among others, the USSR has been able to obtain bases or well-sited facilities: 29 Soviet vessels now operate in the Indian Ocean.

Today, Suez and the Red Sea have ceased to be the main communication route to Europe and the Atlantic. But the Strait of Hormuz remains vital. Since 1967, tankers have carried oil by the Cape route—which enhances the already great strategic importance of southern Africa, where the Soviet Union gained a foothold in 1975 in Mozambique.

No coastal state, not even Australia or India, has sufficient sea power to play a role. The United States remains, with the support notably of Great Britain and France, the guarantor of the status quo.

Toward eastern Asia and Japan, ships use the two traditional routes through the Strait of Malacca and the Strait of Sunda* and, increasingly, the Strait of Lombok, because of its depth.

---

*The distance from the Persian Gulf to Japan is 6,500 miles by the first route and 7,500 miles by the second.

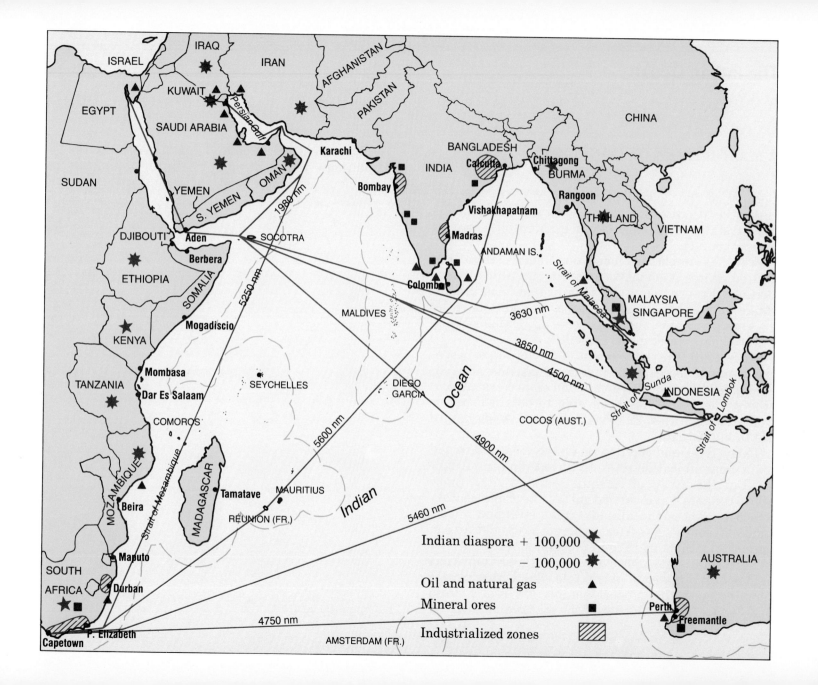

ISRAEL
IRAQ
IRAN
EGYPT
KUWAIT
AFGHANISTAN
SAUDI ARABIA
Persian Gulf
PAKISTAN
CHINA
Karachi
BANGLADESH
SUDAN
YEMEN
OMAN
S. YEMEN
INDIA
Calcutta
Chittagong
BURMA
Bombay
Rangoon
DJIBOUTI
Aden
SOCOTRA
1980 nm
Vishakhapatnam
THAILAND
VIETNAM
Berbera
ETHIOPIA
Madras
5250 nm
ANDAMAN IS.
Strait of Malacca
MALAYSIA
SINGAPORE
SOMALIA
Colombo
3630 nm
KENYA
Mogadiscio
MALDIVES
3850 nm
Mombasa
SEYCHELLES
4500 nm
Strait of Sunda
INDONESIA
TANZANIA
Dar Es Salaam
Ocean
DIEGO GARCIA
Strait of Lombok
COMOROS
5600 nm
COCOS (AUST.)
4900 nm
MOZAMBIQUE
Strait of Mozambique
MADAGASCAR
Tamatave
MAURITIUS
Indian
Beira
RÉUNION (FR.)
5460 nm
AUSTRALIA
Maputo
SOUTH
AFRICA
Durban
Indian diaspora + 100,000
− 100,000
Perth
Freemantle
P. Elizabeth
Oil and natural gas
Capetown
AMSTERDAM (FR.)
4750 nm
Mineral ores
Industrialized zones

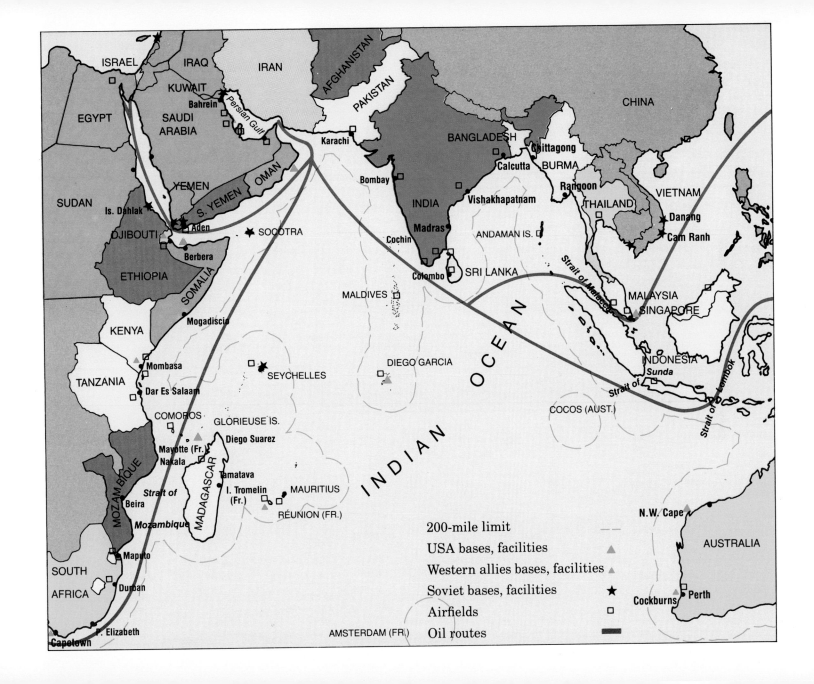

ISRAEL
IRAQ
IRAN
AFGHANISTAN
KUWAIT
EGYPT
Bahrein
SAUDI
ARABIA
PAKISTAN
CHINA
Persian Gulf
Karachi
BANGLADESH
YEMEN
OMAN
Chittagong
Calcutta
BURMA
SUDAN
S. YEMEN
Bombay
Rangoon
VIETNAM
Is. Dahlak
Aden
Vishakhapatnam
THAILAND
Danang
DJIBOUTI
Berbera
SOCOTRA
INDIA
Madras
Cam Ranh
ETHIOPIA
SOMALIA
Cochin
ANDAMAN IS.
Colombo
MALAYSIA
KENYA
Mogadiscio
SRI LANKA
Strait of Malacca
SINGAPORE
MALDIVES
Mombasa
DIEGO GARCIA
INDONESIA
TANZANIA
SEYCHELLES
Sunda
Dar Es Salaam
Strait of
COMOROS
GLORIEUSE IS.
COCOS (AUST.)
Strait of Lombok
Mayotte (Fr.)
Diego Suarez
Nakala
Tamatava
MADAGASCAR
I. Tromelin
(Fr.)
MAURITIUS
N.W. Cape
MOZAMBIQUE
RÉUNION (FR.)
Beira
Strait of
AUSTRALIA
Mozambique
Maputo
SOUTH
AFRICA
Durban
P. Elizabeth
Cockburns
Perth
Capetown

INDIAN OCEAN

200-mile limit
USA bases, facilities
Western allies bases, facilities
Soviet bases, facilities
Airfields
AMSTERDAM (FR.)
Oil routes

The Persian Gulf and the Strait of Hormuz

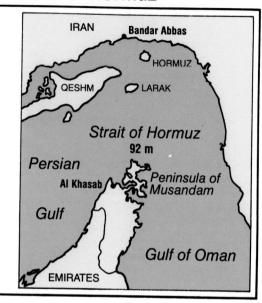

Map: The Persian Gulf and the Strait of Hormuz

IRAQ — Basra, Abadan, Bandar e Mashur, Khorramshahr, Umm Qasr

KUWAIT — BUBIYAN

IRAN — Busheher, Bandar Abbas

Persian Gulf

SAUDI ARABIA — Ras Tannurah, Ad Dammam, BAHRAIN

QATAR — Doha

LAVAN, QEYS, QESHM

Strait of Hormuz

Dubai, Abu Dhabi

UNITED ARAB EMIRATES — Fujairah, Sohar

OMAN

Indian Ocean

Inset map:
IRAN — Bandar Abbas, HORMUZ, LARAK, QESHM
Strait of Hormuz 92 m
Persian Gulf — Al Khasab
Peninsula of Musandam
EMIRATES
Gulf of Oman

Outlets from the Indian Ocean

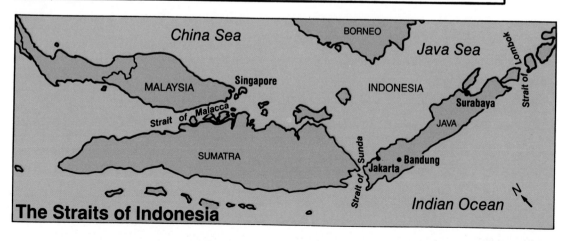

The Straits of Indonesia

China Sea, BORNEO, Java Sea

MALAYSIA — Singapore, Strait of Malacca

SUMATRA

INDONESIA — Surabaya, JAVA, Jakarta, Bandung

Strait of Sunda, Strait of Lombok

Indian Ocean

N

THE RED SEA

Mediterranean

ISRAEL
Tel Aviv
Jerusalem
Amman

Alexandria
Cairo
Port Said
Ismailiya
Suez

JORDAN

Elat

Tabuk

EGYPT

SAUDI ARABIA

Lake Aswan

Red

Sea

Jiddah
Mecca

SUDAN

Dahlak

Asmera

Sana

YEMEN

S. YEMEN

Assab

Aden

ETHIOPIA

Djibouti

Gulf of
Aden

Berbera

The Outlet toward Europe: The Red Sea

THE STRAIT OF TIRAN

Gulf of
Aqaba

SAUDI
ARABIA

SINAI

Strait of Tiran

TIRAN

SINAFIR

Sharm el Sheikh

Red Sea

Red Sea

YEMEN

ETHIOPIA

S.
YEMEN

Bab el
Mandab

PERIM

Gulf of
Aden

DJIBOUTI

DJEZIRET SOBA

THE STRAIT OF BAB EL MANDEB

THE SUEZ CANAL

Mediterranean

MANZALA

Port Said

El Kamtara

Ismailiya

TIMSAH

GREAT
BITTER LAKE

LITTLE
BITTER
LAKE

Suez

Red Sea

Ain Sukhna

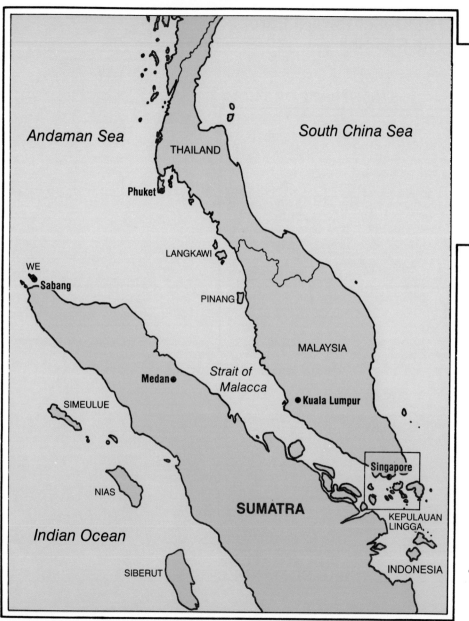

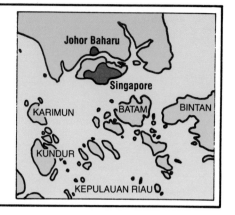

Singapore

Singapore remains one of the key points in Southeast Asia, both strategically and economically.

The Strait of Malacca

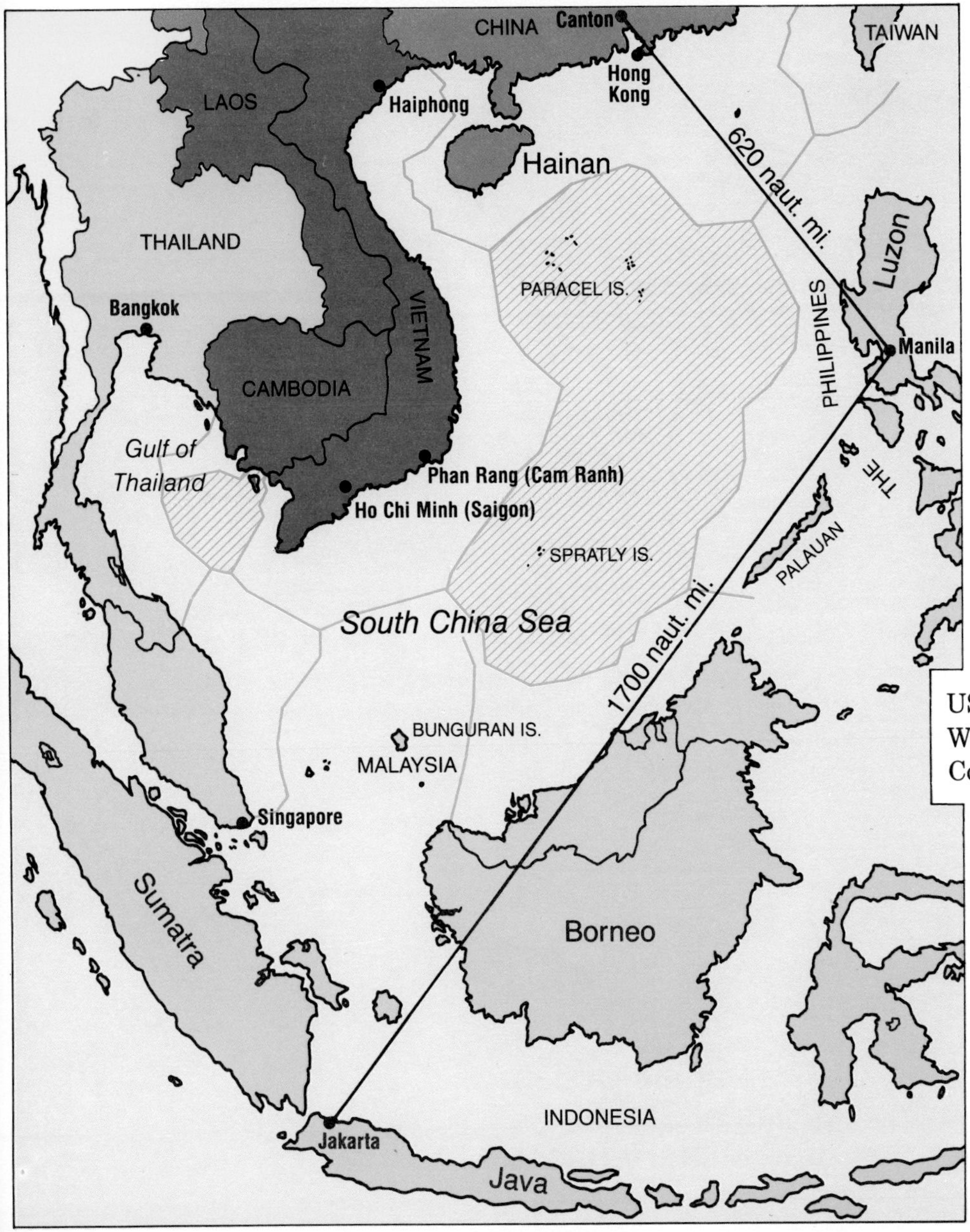

Disputed Areas in the South China Sea

In this region, there remain unresolved problems of sovereignty over parts of the continental shelf. In the northeast, the Chinese and Vietnamese are in dispute over the Paracel Islands, which have been occupied by the Chinese since the mid-1970s. In the south, Indonesia, the Philippines, Taiwan, and Vietnam all seek to assert their sovereignty over the Spratly Islands. Vietnam and Indonesia are in dispute over the continental shelf. In the Gulf of Thailand, Vietnam, Cambodia, and Thailand are in dispute.

CHINA · Canton
TAIWAN
Hong Kong
LAOS · Haiphong
Hainan
620 naut. mi.
THAILAND
Luzon
PARACEL IS.
PHILIPPINES
· Bangkok
VIETNAM
Manila
CAMBODIA
· Phan Rang (Cam Ranh)
Gulf of Thailand
· Ho Chi Minh (Saigon)
THE
PALAUAN
SPRATLY IS.
South China Sea
1700 naut. mi.
BUNGURAN IS.
MALAYSIA
· Singapore
Borneo
Sumatra
INDONESIA
· Jakarta
Java

USSR and pro-Soviet ■
Western allies ■
Contested waters ▨

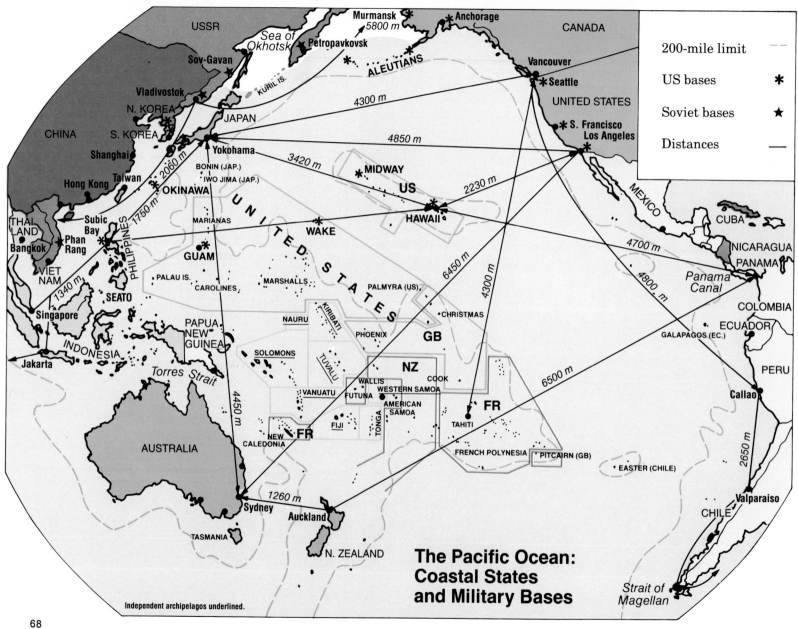

The Pacific Ocean: Coastal States and Military Bases

Independent archipelagos underlined.

Legend:
- 200-mile limit — — —
- US bases ✳
- Soviet bases ★
- Distances ——

68

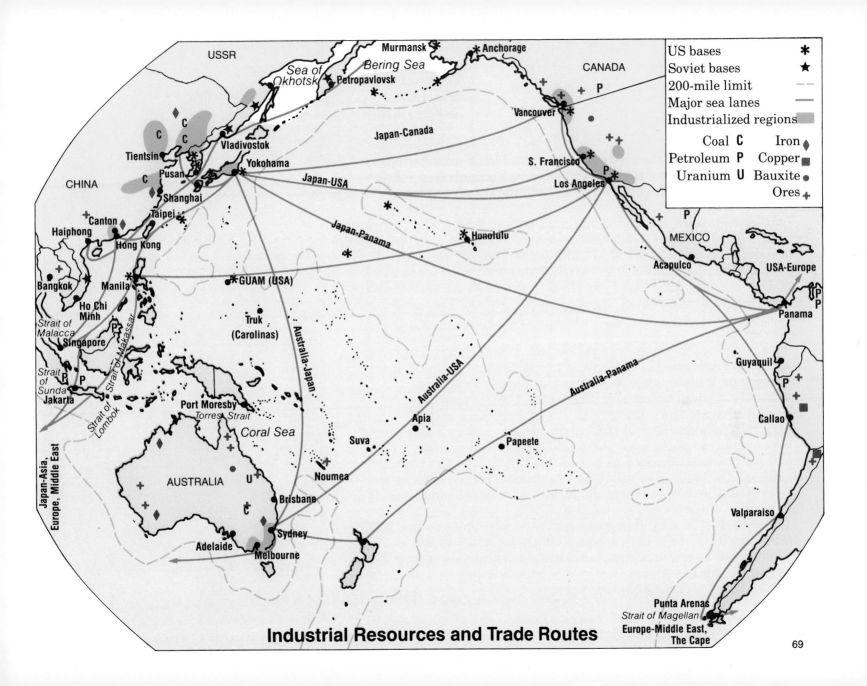

Industrial Resources and Trade Routes

Legend	
US bases	✳
Soviet bases	★
200-mile limit	- - -
Major sea lanes	——
Industrialized regions	▮

Coal **C**   Iron ◆
Petroleum **P**   Copper ■
Uranium **U**   Bauxite ●
Ores ✚

USSR

Sea of Okhotsk

Bering Sea

Murmansk ✳   ✳ Anchorage

CANADA

Petropavlovsk ★

Japan-Canada

Vancouver ✳   P

✚   ✚

Vladivostok ★

Tientsin

C   C   C

Yokohama ✳

S. Francisco ✳

Japan-USA

Los Angeles ✳   P   ✳

Pusan
Shanghai
CHINA
C

Taipei ✳

Japan-Panama

MEXICO

Canton ✚

Haiphong

Hong Kong

✳ Honolulu

Acapulco

USA-Europe

Bangkok

Manila ✳

GUAM (USA) ✳

P   P
Panama

Ho Chi Minh

Strait of Malacca

Truk (Carolinas)

Australia-Japan

Guyaquil

Singapore   P

Strait of Makassar

Australia-USA

Australia-Panama

P

Strait of Sunda   P   P
Jakarta

Strait of Lombok

Port Moresby
Torres Strait

Coral Sea

Apia

Papeete

Callao ■

Suva

Japan-Asia, Europe, Middle East

Noumea

AUSTRALIA   U ✚

Brisbane

C   ◆

Valparaiso ■

Adelaide

Sydney

Melbourne

Punta Arenas
Strait of Magellan

Europe-Middle East, The Cape

69

The Pacific Ocean

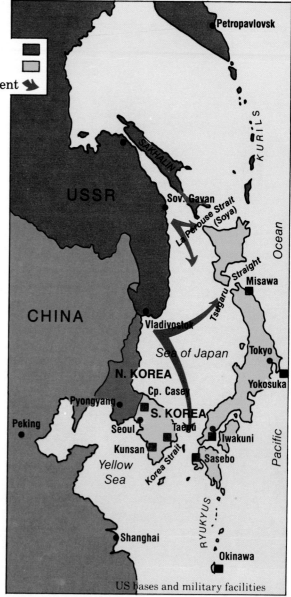

USSR and pro-Soviet ■
Western allies ■
Soviet naval deployment �’

Before World War II, the great powers showed little interest in this vast ocean. Their maritime activities were concentrated for the most part in the southwestern Pacific, and the ocean was merely a secondary highway.

With its seaboard on the Pacific, only the United States has, since the nineteenth century, pursued a coherent expansionist policy through the archipelagos to the Philippines. During World War II, Japan saw its imperial plans thwarted. Today, under the pressure of political changes that have occurred since 1945 in China and Southeast Asia, the role of the Pacific has grown. The economic importance of Japan and the growth of various states have intensified trade flows.

However, the Pacific has not yet become a new center of gravity, equal to the Atlantic. The continued preponderance of the Atlantic derives from the economic importance of Europe and the fact that both sides of that ocean are peopled by those sharing common origins and a common culture.

Control of Sakhalin and the Kuril Islands (formerly Japanese possessions) is important for the USSR. On its Pacific coast, only a small part of which is open to navigation throughout the year, the Japanese straits form so many gates opposite Vladivostok. Petropavlovsk, at the extreme tip of the Kamchatka, is icebound three or four months of the year.

In the Pacific, the USSR had no base until 1979. Since then, it has used Cam Ranh Bay in Vietnam, which is both a base and a communications center. Naval—and air—facilities also exist at Da Nang in Vietnam (see page 67).

The Japanese Gates

US bases and military facilities

Hawaii: Key Position in the Pacific

Most of the islands in the Pacific are under the trusteeship of the West and its allies (USA, Great Britain, France, New Zealand, Australia, Japan). U.S. hegemony is almost total in the northern Pacific. The central position of Hawaii, as base and as staging area, is vital in this arrangement.

The archipelago contains the major base of Pearl Harbor, which acts as support base for the Pacific fleet, and Camp Nimitz, the seat of the Armed Forces Pacific Command, and it is also a major air traffic center.

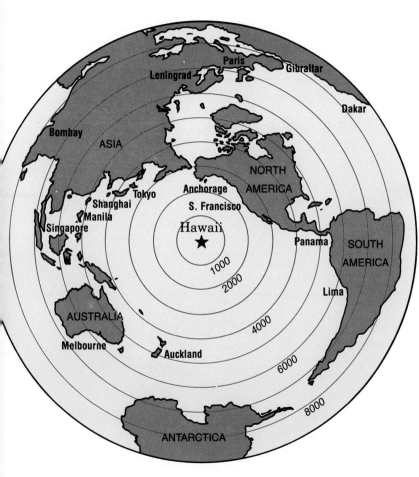

Hawaii: The Fiftieth American State

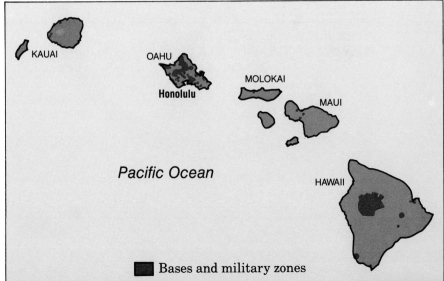

■ Bases and military zones

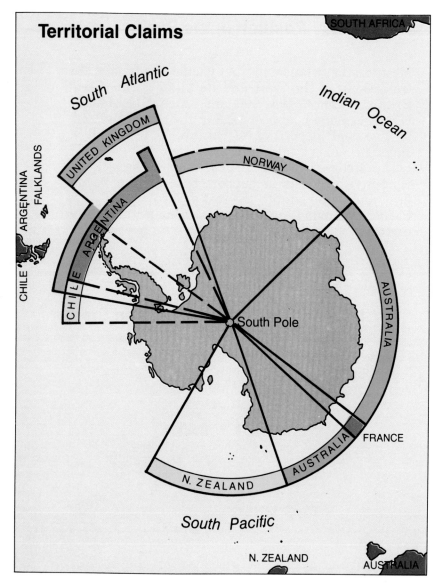

Territorial Claims

South Atlantic

Indian Ocean

UNITED KINGDOM

NORWAY

ARGENTINA

CHILE

ARGENTINA

AUSTRALIA

CHILE

ARGENTINA

FALKLANDS

South Pole

FRANCE

AUSTRALIA

N. ZEALAND

South Pacific

N. ZEALAND

AUSTRALIA

SOUTH AFRICA

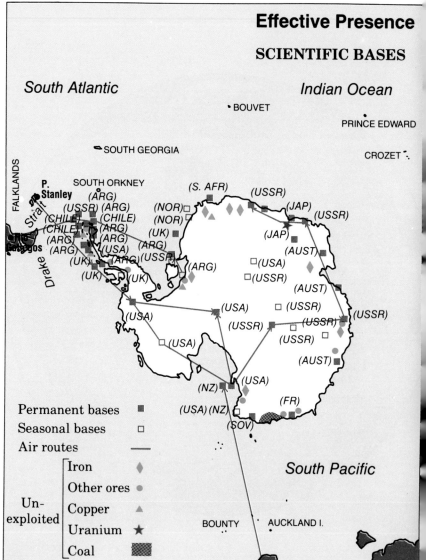

Effective Presence

SCIENTIFIC BASES

South Atlantic

Indian Ocean

BOUVET

PRINCE EDWARD

SOUTH GEORGIA

CROZET

FALKLANDS

P. Stanley

SOUTH ORKNEY

(S. AFR)

(USSR)

(ARG)

(JAP)

(USSR)

(USSR)

(ARG)

(NOR)

(NOR)

(JAP)

Drake Strait

(CHILE)

(CHILE)

(CHILE)

(ARG)

(UK)

(AUST)

(ARG)

(ARG)

(ARG)

(USA)

(ARG)

(USSR)

(USA)

(USSR)

Galapagos

(ARG)

(USA)

(USSR)

(UK)

(ARG)

(USSR)

(AUST)

(UK)

(UK)

(USA)

(USSR)

(USSR)

(USA)

(USSR)

(USA)

(USSR)

(USSR)

(AUST)

(NZ)

(USA)

(USA) (NZ)

(FR)

(SOV)

Permanent bases ■

Seasonal bases ☐

Air routes ———

Un-exploited
- ◆ Iron
- ● Other ores
- ▲ Copper
- ★ Uranium
- ▦ Coal

South Pacific

BOUNTY   AUCKLAND I.

Antarctica

Since 1900, seven states have claimed territorial rights over portions of the Antarctic continent:

United Kingdom (1908)
New Zealand (1923)
France (1924 and 1938)
Australia (1933)
Norway (1939)
Chile (1940)
Argentina (1943)

More recently, Brazil has added itself to this list.

This partition is largely formal, since the United States, the USSR, and other states have always reserved their claims.

Brazil, Chile, and Argentina base their claims on geographic proximity, the latter two disputing rights asserted by the British. The British, by maintaining their sovereignty over the Falkland Islands in 1982, continue to exercise de facto authority over the sector they laid claim to in 1908.

On Antarctica itself, several countries have an effective presence, permanent or seasonal, in the form of meteorological or scientific stations: in quantitative order, the USSR (12), the USA (9), Argentina (8), Australia (3), Japan (2), New Zealand (3), United Kingdom (4), France (1).

The Falkland Islands and Antarctica

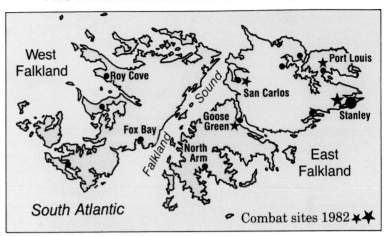

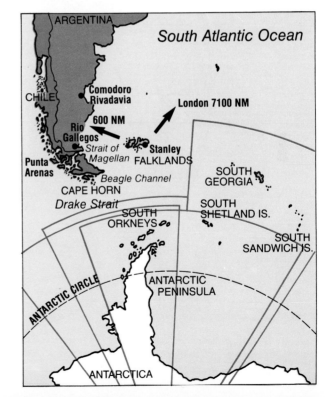

73

The World from Antarctica

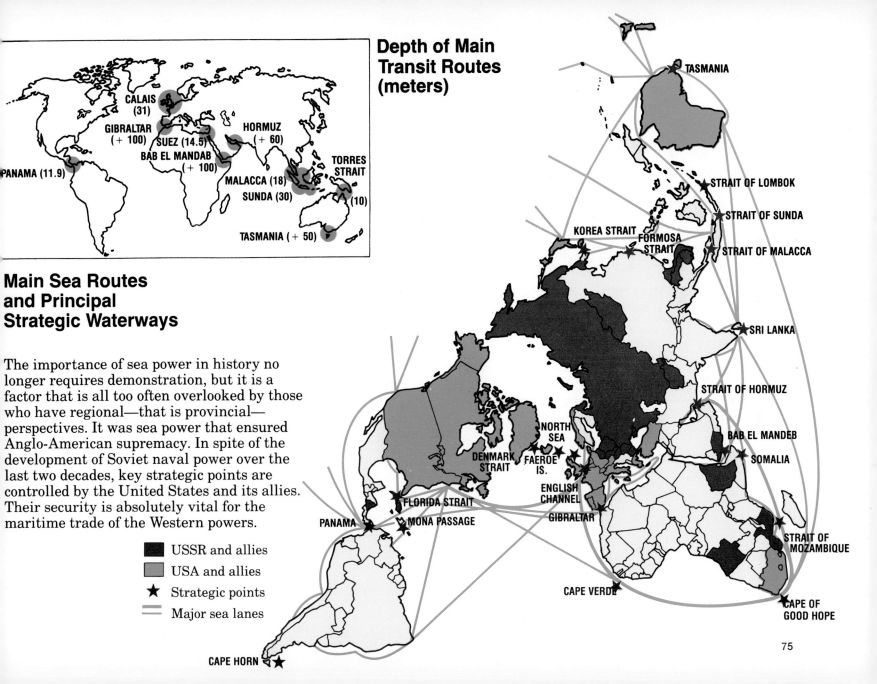

Depth of Main Transit Routes (meters)

CALAIS (31)
GIBRALTAR (+ 100)
SUEZ (14.5)
BAB EL MANDAB (+ 100)
HORMUZ (+ 60)
PANAMA (11.9)
MALACCA (18)
SUNDA (30)
TORRES STRAIT (10)
TASMANIA (+ 50)

Main Sea Routes and Principal Strategic Waterways

The importance of sea power in history no longer requires demonstration, but it is a factor that is all too often overlooked by those who have regional—that is provincial—perspectives. It was sea power that ensured Anglo-American supremacy. In spite of the development of Soviet naval power over the last two decades, key strategic points are controlled by the United States and its allies. Their security is absolutely vital for the maritime trade of the Western powers.

USSR and allies
USA and allies
★ Strategic points
Major sea lanes

TASMANIA
STRAIT OF LOMBOK
STRAIT OF SUNDA
STRAIT OF MALACCA
KOREA STRAIT
FORMOSA STRAIT
SRI LANKA
STRAIT OF HORMUZ
BAB EL MANDEB
SOMALIA
NORTH SEA
DENMARK STRAIT
FAEROE IS.
ENGLISH CHANNEL
GIBRALTAR
FLORIDA STRAIT
MONA PASSAGE
PANAMA
STRAIT OF MOZAMBIQUE
CAPE VERDE
CAPE OF GOOD HOPE
CAPE HORN

SECURITY PERCEPTIONS OF THE UNITED STATES, THE USSR, AND REGIONAL AND MIDDLE POWERS

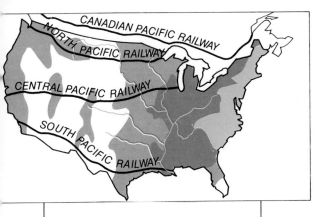

How the United States Was Settled

Occupied in 1800
Occupied about 1850
Occupied about 1900

The United States

1. Delaware
2. New Jersey
3. Rhode Island
4. Massachusetts
5. New Hampshire
6. Connecticut
7. Vermont
8. Maryland

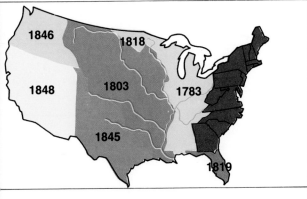

1846
1818
1848
1803
1783
1845
1819

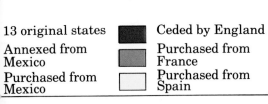

13 original states

Annexed from Mexico

Purchased from Mexico

Ceded by England

Purchased from France

Purchased from Spain

Territorial Growth

The United States:
Formation and Settlement

In the first phase of its expansion, the United States was able, without fighting any major wars, rapidly to occupy its hinterland. It had no rivals on the continent. It was soon making its presence felt outside the continent (Liberia, 1820; Japan, 1854; China, 1859), and by the turn of the century had ensured its economic dominance over the American continent and occupied the Pacific as far as the Philippines. Until the Korean war, the USA had known only absolute victories, total successes: the conquest of the West, the purchase of land from foreign powers (Alaska in 1867), wars against Mexico and Spain. The interventions in 1917 and 1941 themselves were wars in which the United States was protected by its geographic isolation.

As a business society with an empirical philosophy, a Protestant ethic, and a liberal democracy with great social mobility, the United States, proud of having neither colonies nor rivals, had not throughout its history developed any real conception of interstate relations. Not having known the realities that spring from a constant struggle for national self-preservation, the United States became a world power with a historical experience very different from that of the European states.

After World War II, the United States, the dominant state on the planet, found itself faced with the task of containing Communist expansion (1947).

Strong in its nuclear monopoly, then enjoying an undeniable superiority for two decades, the United States sought first and foremost to contain this expansion systematically.

The American withdrawal, noticeable after the war in Vietnam (1973), which profoundly affected public opinion, and consequently American political will, is to be traced to a number of factors: conceptual weakness* and lack of knowledge of the outside world; frequent confusion of national aspirations with the expansion of communism; defense of notoriously corrupt and ineffective allies solely because they are pliable, combined with an inability to support allies that have been promised security. As for decisions, they are generally made in moments of crisis, while long-term planning is neglected.

However, by reintroducing human rights as a political point of reference, the United States launched the first Western ideological counteroffensive in four decades, not to mention, more recently, an ambitious rearmament program intended to reassert its superiority.

Whatever its economic difficulties, the United States possesses the necessary technological advances, the resources, and the dynamism to continue to be the leading power in the world.

---

*The Nixon–Kissinger team is an exception.

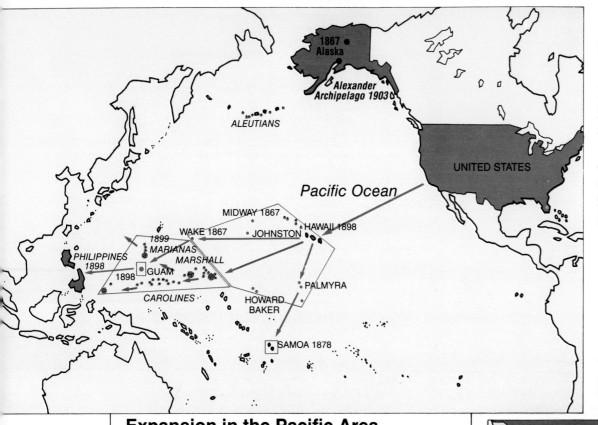

Expansion in the Pacific

1945—Micronesia (Marianas, Carolines, Marshalls, Western Samoa)

The Philippines, an American colony since 1898, became independent in 1945.

1867 ● Alaska

Alexander Archipelago 1903

ALEUTIANS

UNITED STATES

Pacific Ocean

MIDWAY 1867
WAKE 1867
JOHNSTON
HAWAII 1898
1899 MARIANAS
MARSHALL
PHILIPPINES 1898
1898 GUAM
CAROLINES
PALMYRA
HOWARD BAKER
SAMOA 1878

Expansion in the Pacific Area

In 1945 ▮   In the 19th century ▮

Interventions in the Caribbean Basin

There have been some fifteen major interventions in the Caribbean basin between 1903 (Panama) and 1965 (Dominican Republic)

The United States in Central America

MEXICO
1914 1916
1898-1906 1917
1921-1933
CUBA
PUERTO RICO
1898
SANTO DOMINGO
1905 1916-1965
HAITI
1912 1915
NICARAGUA
1909
PANAMA 1903

Protectorates ▨   US interventions ←
Other countries ☐   USA and annexations ▮

Minorities in the United States

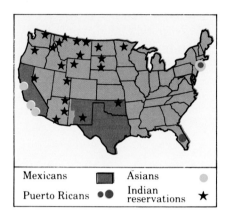

Mexicans ▨
Puerto Ricans ●●
Asians ●
Indian reservations ★

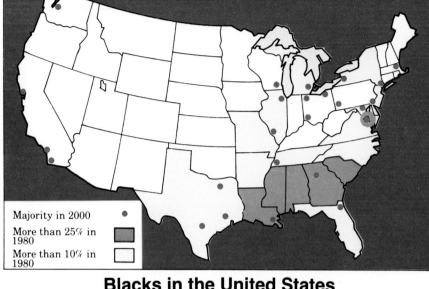

Majority in 2000 ●
More than 25% in 1980 ▨
More than 10% in 1980 □

Blacks in the United States

While the Indian minority forms only a tiny percentage of the population, blacks and Hispanics (such as the Chicanos or the Puerto Ricans) number some 50 million individuals: blacks, 26.5 million; Hispanics, 14.5 million (1980). In addition there are 8 million illegal aliens.

With Asians* and other minorities, the United States population is 20 to 22 percent "nonwhite." Blacks, increasingly numerous in the metropolitan areas of the Northeast and the West (New York, Philadelphia, Washington, Detroit, Chicago, Los Angeles, San Diego, Seattle), are also concentrated in several cities in the South (Houston, Dallas, Memphis, Atlanta). Hispanics are mainly settled in California, New Mexico, and Texas, and in the following

cities: Los Angeles, New York, El Paso, Miami, San Antonio.

It is estimated that, since 1975, legally or otherwise, about 1 million immigrants have entered the United States each year, 82 percent of them from Latin America or Asia. Over the last decade, Los Angeles has come to play the role formerly held by New York as the mecca for immigrants from all over the world. The combination of poverty and racial tensions, particularly among blacks, intensifies the feeling of insecurity in the big cities.

---

*The Asian communities (Japanese, Chinese, Koreans, and so on) are often well-to-do.

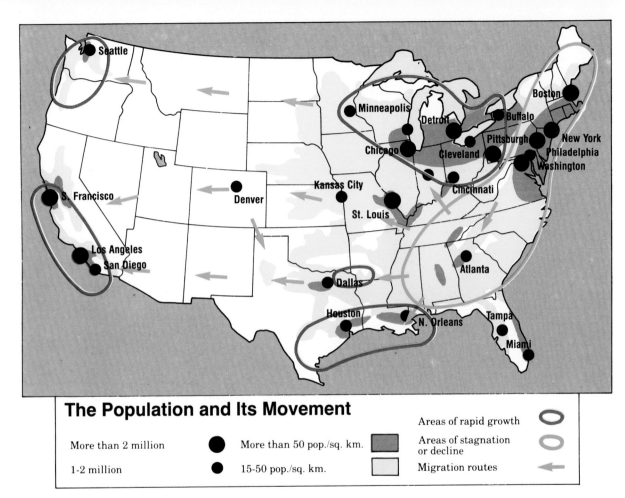

The Population and Its Movement

More than 2 million ●

1-2 million ●

More than 50 pop./sq. km. ●

15-50 pop./sq. km. ●

Areas of rapid growth ⬭

Areas of stagnation or decline ⬭

Migration routes ←

United States: Internal Migrations

The old industrial East, bordered by the Great Lakes–Washington–Boston triangle, is increasingly being replaced, in terms of resources and soon in terms of population, by the West, especially California and Texas.

Migration movements, both internal and external (coming mainly from Mexico and the Caribbean), are toward the southern and western parts of the country. A gradual movement of the country's center of gravity is thus under way.

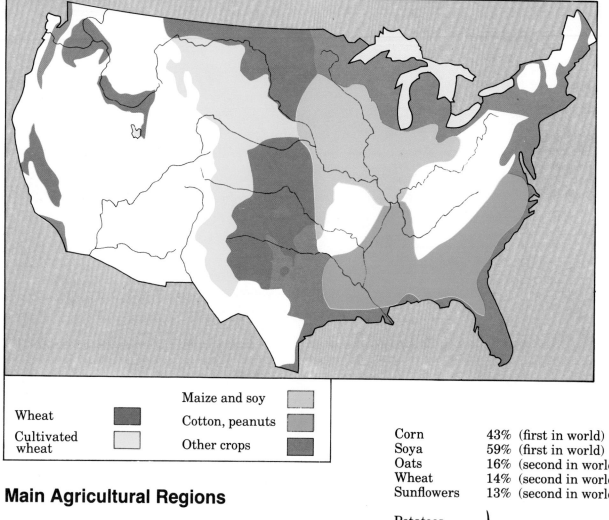

Wheat	■	Maize and soy		□
Cultivated wheat	□	Cotton, peanuts		■
		Other crops		■

Main Agricultural Regions

Thanks to a remarkable productivity, the United States is the leading agricultural producer in the world. It is far and away the leading exporter of grains and soya products.

Corn	43%	(first in world)
Soya	59%	(first in world)
Oats	16%	(second in world)
Wheat	14%	(second in world)
Sunflowers	13%	(second in world)
Potatoes		
Sugar beets		
Peanuts	}	(third in world)
Beef		
Pork		

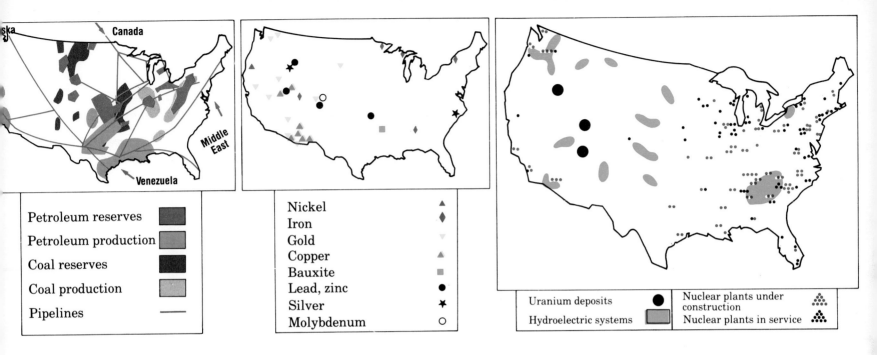

Petroleum reserves
Petroleum production
Coal reserves
Coal production
Pipelines

Nickel
Iron
Gold
Copper
Bauxite
Lead, zinc
Silver
Molybdenum

Uranium deposits
Hydroelectric systems
Nuclear plants under construction
Nuclear plants in service

United States: Resources

With substantial energy and mineral resources, the United States remains the world's leading industrial power.

In leading sectors such as aerospace, aeronautics, computers, and nuclear development, the United States effectively demonstrates its determination to maintain or increase its lead.

Minerals

Copper	15%	(first in world)
Iron	8%	(third in world)
Lead	15%	(second in world)
Zinc	6%	(fifth in world)
Silver	10%	(fourth in world)

Gold, nickel, bauxite, etc.

Energy Resources

Coal	25%	(first in world)
Uranium	36%	(first in world)
Natural gas	35%	(first in world)
Petroleum	15%	(third in world)
Hydroelectricity		(first in world)

Industrial Centers

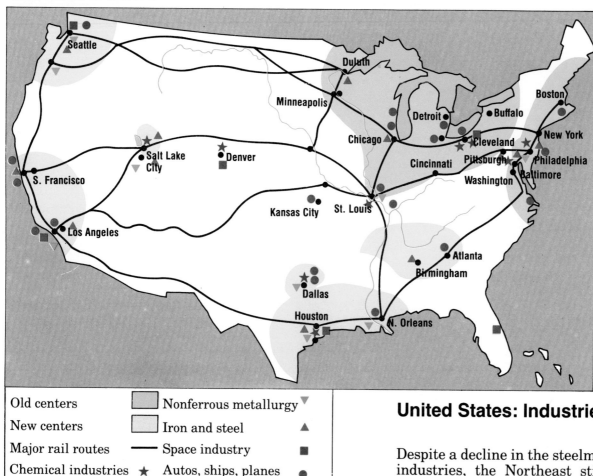

Legend:

Old centers
New centers
Major rail routes
Chemical industries ★

Nonferrous metallurgy ▼
Iron and steel ▲
— Space industry ■
Autos, ships, planes ●

Information industry: the Northeast, Texas

Map labels: Seattle, Duluth, Minneapolis, Detroit, Buffalo, Boston, Chicago, Cleveland, New York, Cincinnati, Pittsburgh, Philadelphia, Salt Lake City, Denver, Washington, Baltimore, S. Francisco, Kansas City, St. Louis, Los Angeles, Atlanta, Birmingham, Dallas, Houston, N. Orleans

United States: Industries

Despite a decline in the steelmaking and engineering industries, the Northeast still remains the traditional stronghold of heavy industry.

The South (particularly Texas), the Pacific coast, and more recently the Rockies (around Salt Lake City and Denver) are undergoing rapid development thanks to new industrial sectors (nuclear energy, aeronautics, chemistry).

A RESERVOIR OF RESOURCES

Situated between the United States and Alaska, Canada is marked by its vast size, its continental climate with an exceptionally harsh winter, and its very small population. Settlement quite closely follows the line of the border with the United States, although the country also has an Arctic inclination. The economy is largely integrated into the United States market.

Does such a situation encourage the building of a nation? Paradoxically, the particularism of French-speaking Quebec, involving a degree of linguistic and cultural opposition to the United States and Anglo-American hegemony, may contribute to nationhood.

Agricultural Regions

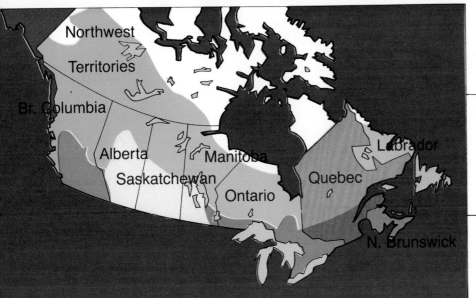

Northwest Territories
Br. Columbia
Alberta
Saskatchewan
Manitoba
Ontario
Quebec
Labrador
N. Brunswick

- French-speaking Quebec
- Forests
- Various crops
- Grain regions

A Rich Subsoil

Dawson
Uranium City
Churchill
Edmonton
Vancouver
Calgary
Regina
Winnipeg
Quebec
Montreal
Ottawa
Toronto

- Petroleum
- Pipelines
- Coal
- Iron
- Uranium
- Copper
- Silver
- Gold
- Hydroelectricity

Energy and Mineral Resources

Petroleum	3%	(w.p.) res.*
Gas	4.5	(w.p.) res.
Uranium	16	(w.p.) res. 19%
Hydroelectricity	16	(w.p.)
Iron	5.5	(w.p.) res. 8.5%

*(w.p.) res. = (world production) reserves.

USSR:
Historical Expansion

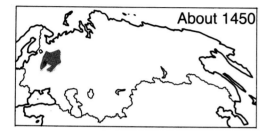

About 1450

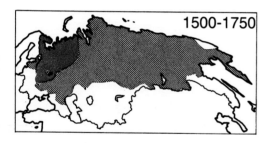

1500-1750

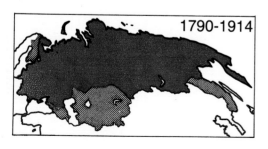

1790-1914

Long occupied by the Mongols (thirteenth through fifteenth centuries), Russia has undergone two processes: The first, uninterrupted since the sixteenth century, was the conquest of its hinterland, followed by imperial conquests at the expense particularly of the Ottoman and Chinese empires; the other was the pursuit of Europeanization to overcome its backwardness.

If we ignore Leninism and socialist ideology, we see that the Soviet Union has retained and increased the imperial heritage in new forms. Since the beginning of the five-year plans, it has been attempting to make up for its industrial backwardness and to develop its military capacities.

In 1948 Finland and Yugoslavia left the Soviet sphere of influence.

Expansion from 1938 to 1950

Pelsamo
FINLAND
Vyborg
Leningrad
Baltic lands
USSR
Koenigsberg
Vilno
FRG
POLAND
Brest-Litovsk
CZECHOSLOVAKIA
Lvov
Ruthenia
HUNGARY
Bessarabia
YUGOSLAVIA
ROMANIA
BULGARIA
ALBANIA

USSR in 1938	
Annexations	
Sphere of Influence	

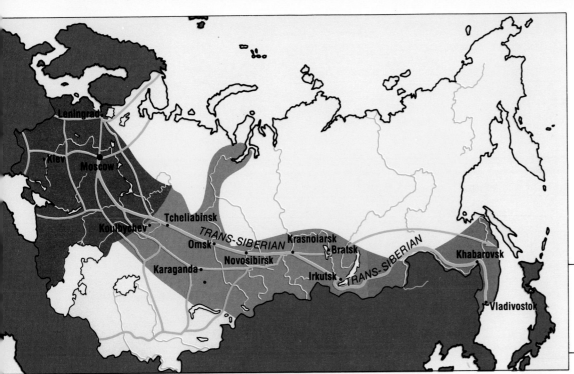

The Conquest of the East

Stages in the
occupation of the land

Before 1920	
1920-1955	
After 1955	
Railways	

USSR

Although tsarist Russia conquered Siberia over the last few centuries and built the first Trans-Siberian railway, significant settlement, the development of communications, and industrialization date from the last five decades. Southern Siberia has been settled mostly by Slavs. In this way, despite ideological rivalries, Europe has reached the Pacific. American conquest of its Far West corresponds, making all due allowances, to the Russian conquest of the East.

Expansion of the Slavs in the USSR

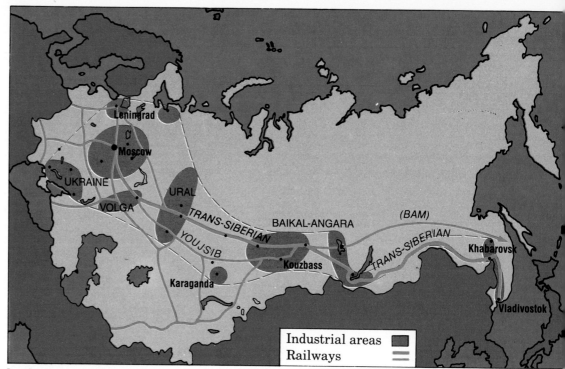

Industrial Expansion Eastward

Industrial areas
Railways

Leningrad
Moscow
UKRAINE
VOLGA
URAL
TRANS-SIBERIAN
YOU JSIB
BAIKAL-ANGARA
(BAM)
TRANS-SIBERIAN
Kouzbass
Karaganda
Khabarovsk
Vladivostok

Soviet Strategic Bases

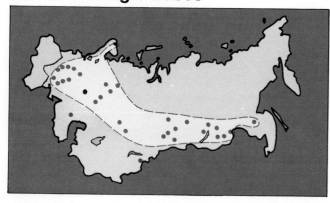

USSR

The nuclear sites stretch through Soviet territory along a line that closely follows the area of Slav settlement.

Kazakhstan, which is the only Muslim republic to have major nuclear and aerospace sites, has a population that is 43 percent Slav.

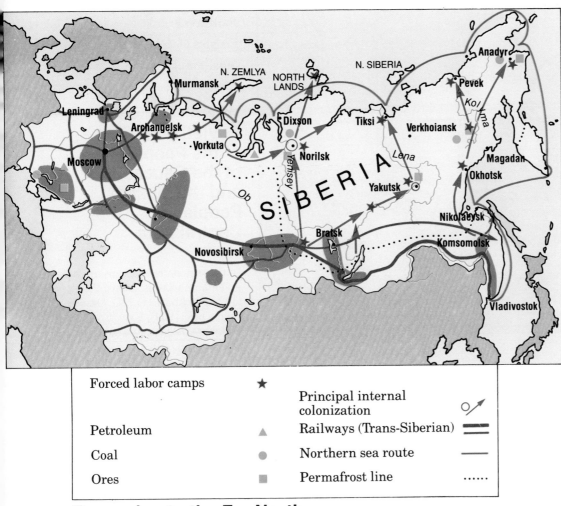

USSR

Moving out from the Trans-Siberian railway, the gradual and arduous conquest of northern Siberia has been carried out along the great rivers (Ob, Yenisey, Lena, Kolyma), and along the main sea route linking the White Sea and Vladivostok through the Arctic Ocean. The Soviets use icebreakers that have no equal in the world. It is likely that the Siberian camps have played a not insignificant part in the process of opening up the area.

Forced labor camps	★		
		Principal internal colonization	○⟋➚
Petroleum	▲	Railways (Trans-Siberian)	▬▬
Coal	●	Northern sea route	──
Ores	■	Permafrost line	⋯⋯

Expansion to the Far North

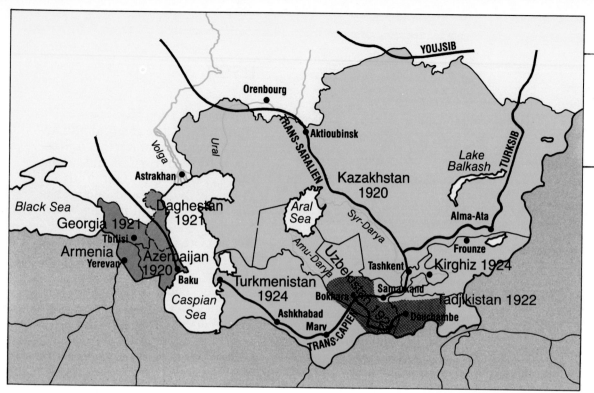

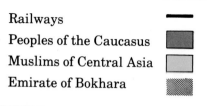

Railways ▬

Peoples of the Caucasus ■

Muslims of Central Asia ■

Emirate of Bokhara ▨

Sovietization of the Peoples of the Caucasus and the Muslim Peoples of Central Asia

In these regions, some groups have been given autonomous status: Cherkesses (1922), Abkhazes (1921), Ossetes (1924), Ingouches (1934), Kara Kalpaks (1925), Ouigours, Adjars, and so on.

USSR

Muslim Central Asia and the Islamic part of the Caucasus were conquered during the nineteenth century, often in the face of fierce resistance.

Today these regions are divided into six republics. This division was established by the Bolshevik government at Stalin's instigation. The Emirate of Bokhara was annexed and divided among several republics in 1920 and was the scene of a prolonged uprising (1920–1928). The nationalities policy seeks to differentiate rather than to unify ethnic groups that are relatively homogeneous (only the Tadjiks are not Turkish-speaking).

The alphabet adopted for the transcription of the languages of these republics is Cyrillic (1928). The population growth rate is distinctly higher than that of the Slav groups. The economic growth of these republics has been spectacular, and Uzbekistan has long been held up as a model for the countries of the Middle East. Georgia and Armenia were sovietized in 1920, after a short-lived independence. These two republics are not Muslim.

The USSR: A Multinational State
The Fifteen Republics and the Principal Autonomous Republics

Frontiers of federal republics	——
Frontiers, autonomous republics	—
Slavic republics	▓
Muslim republics	▓
Caucasian republics	▢
Other republics	▓

In terms of area, the Soviet Union is the largest country in the world (22,000,000 sq. km.), but most of its population lives in the west and south, 70 percent of the population living in the European part on 20 percent of its area. Siberia holds great strategic interest; in addition to its area, it contains vast mineral and forest resources potentially exploitable or already being exploited. But there are only 25 million inhabitants in its 13,000,000 sq. km. On its eastern flank, the USSR controls the Sea of Okhotsk.

While one Soviet citizen in two is not Russian, the other nationalities (officially there are 126 of them) are an extremely diverse group. The two most numerous groups are the Slavs (Ukrainians and Byelorussians)—some 50 million—and the Muslims, mostly Turkish-speaking, who number almost 50 million but who are divided among six republics.

The government of this bureaucratic and totalitarian society rests on the organization provided by the single party, the guarantor of ideological orthodoxy; by the KGB, the guarantor of security; and by the army, the guarantor of Soviet power in the world. The system contains a series of contradictions: agriculture/industry; heavy industry/consumer goods; military sector/civilian sector (marked by bottlenecks, waste, and corruption); national, religious, and political problems. This system, in the framework of its contradictions, engenders passivity, black markets (which often act as regulators), and low productivity.

However, strategically the system benefits from the advantages of its militarized structure, from its secretiveness, from its political cohesion (once goals have been laid down), from the control of its public opinion, and from the role still played by its ideology within and beyond its frontiers. These advantages amply compensate for the ponderous bureaucracy and mitigate the relative lack of dynamism and the shortcomings of the economic machinery.

The USSR is the world's second economic power, just ahead of Japan, and its power is above all military. This military power has been considerably developed and modernized over the last two decades. In a context where the United States was politically paralyzed by the outcome of the Vietnam war, the Soviet Union exploited the vacuums in Angola, Ethiopia, and Afghanistan and demonstrated its logistical capabilities.

The Subsoil of the Soviet Union

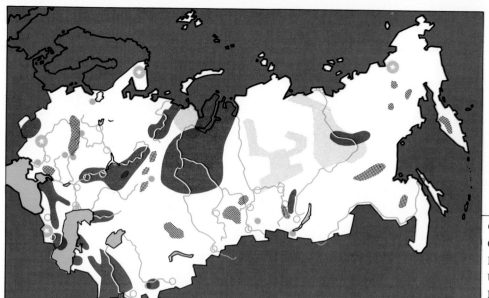

ENERGY RESOURCES

Oil and natural gas
Coal
Hydroelectricity
Uranium
Nuclear energy

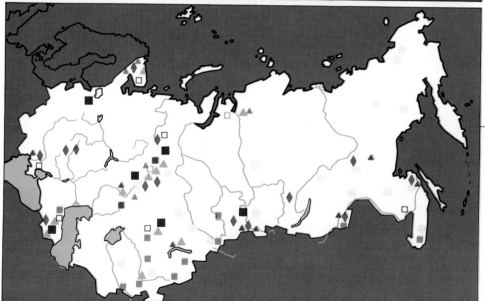

METALS

Platinum
Silver
Chrome
Manganese
Nickel
Lead-Zinc-Tin
Iron
Copper
Gold
Bauxite

70% in Europe on 20% of the territory
30% in Asia on 80% of the territory
126 nationalities, but 15 republics*

Russia	137 million (83% Russian)
The Ukraine	50 million (20% Russian)
Uzbekistan	15 million (12.5% Russian)
Kazakhstan	14.5 million (43% Russian)
Byelorussia	9.5 million (10.5% Russian)
Azerbaijan	5.9 million (10% Russian)
Georgia	5 million (8.5% Russian)
Moldavia	4 million (12% Russian)
Tadjikistan	3.7 million (12% Russian)
Kirghiz Republic	3.5 million (30% Russian)
Lithuania	3.4 million (8.5% Russian)
Armenia	3 million (2.7% Russian)
Turkmenistan	2.7 million (15% Russian)
Latvia	2.5 million (30% Russian)
Estonia	1.5 million (25% Russian)

*126 nationalities are officially recognized. The fifteen with more than a
million inhabitants form the Federated Socialist Soviet Republics.
Depending on their size (in terms of population), the others form
autonomous republics or autonomous territories (notably the nomadic
peoples).

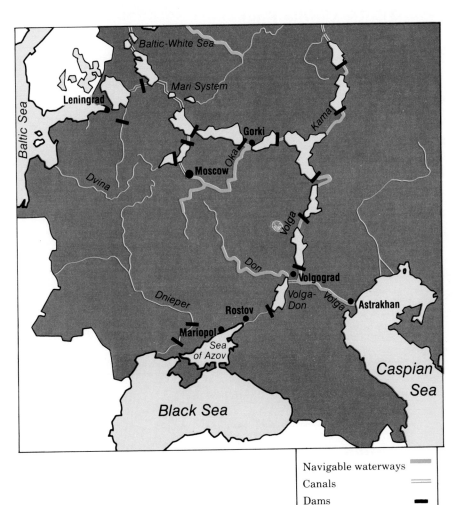

Navigable waterways
Canals
Dams

The North–South Waterway Link

The Soviets have linked up their various operational theaters by means of a network of canals connecting five seas: the Sea of Azov, the Black Sea, the Caspian Sea, the Baltic Sea, and the White Sea. Ships under 5,000 tons can use them.

USSR/Middle East

The maps show Soviet aspirations in the region that they had hoped to control through the Nazi–Soviet Pact (1939).

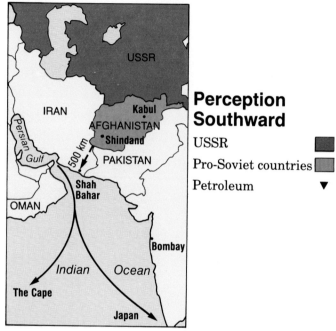

Perception Southward

USSR ▮

Pro-Soviet countries ▨

Petroleum ▼

Afghanistan

The Soviet intervention in Afghanistan (December 1979) brought the Soviets to within 500 km. of the Indian Ocean and underlined the vulnerability of a Pakistan already weakened by its Indian rival.

Coming on top of the fall of the Shah of Iran, this intervention reduced virtually to nil the American position in the Near East.

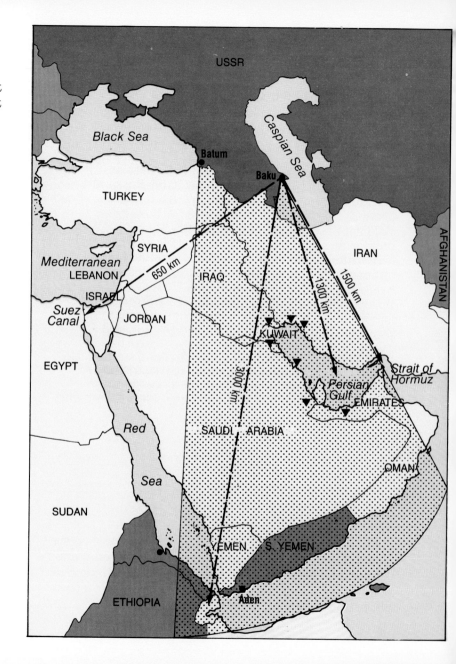

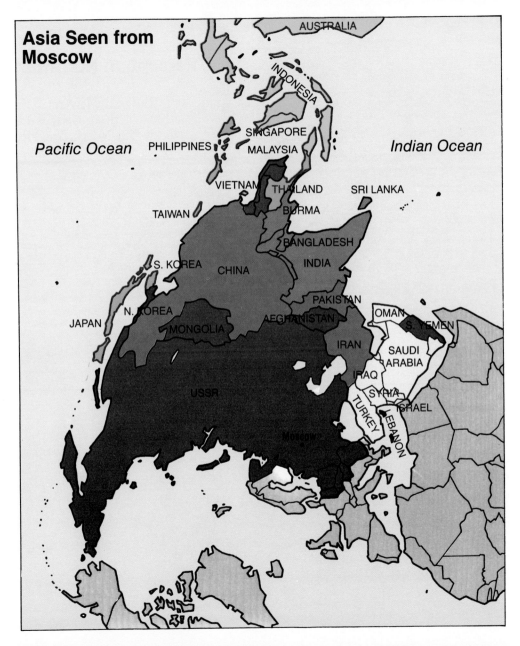

Asia Seen from Moscow

AUSTRALIA

INDONESIA

Pacific Ocean

SINGAPORE

PHILIPPINES

MALAYSIA

Indian Ocean

VIETNAM THAILAND

SRI LANKA

TAIWAN BURMA

BANGLADESH

S. KOREA CHINA INDIA

N. KOREA PAKISTAN

JAPAN AFGHANISTAN OMAN

MONGOLIA S. YEMEN

IRAN SAUDI ARABIA

IRAQ

USSR SYRIA

TURKEY ISRAEL

LEBANON

Moscow

For the USSR, whose northern reaches are in the polar region and which has limited access to the open sea, it is vital to have maritime staging areas for its rapidly growing fleet along the peninsular and insular belt of Asia. Geopolitical logic would require the Soviet Union to have control of the arc running from the Indian subcontinent to the Horn of Africa.

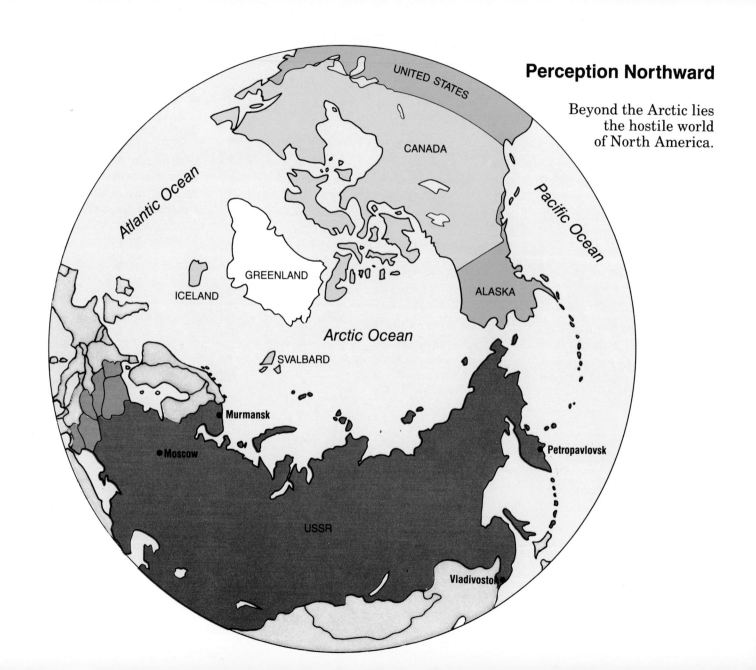

Perception Northward

Beyond the Arctic lies
the hostile world
of North America.

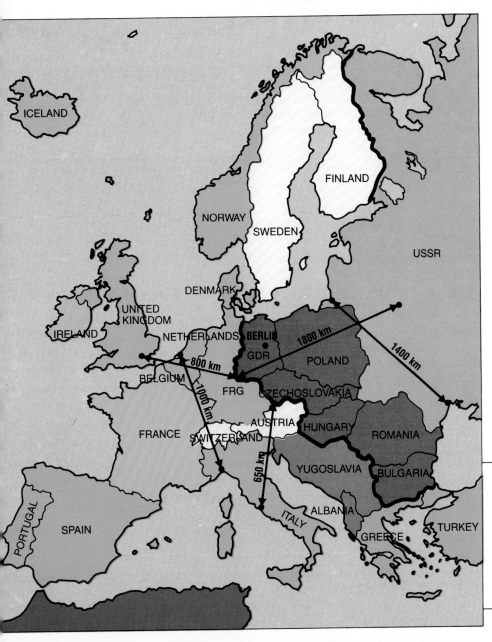

Europe: Two Blocks of Hostile States

Western Europe, the center of the planet until 1918 and the continent on which was played out the future of the world until 1945, is marked today by its political division, its economic power, and its relative military vulnerability.

Whereas the whole of Central Europe fell into the Soviet orbit in 1945, Western Europe, with American support, was able to maintain its overall position. Austria and Finland retained or regained their independence through their neutrality. The only retreats in the Soviet camp, in Yugoslavia (1948) and Albania (1960), were due not to Western pressure, but to the contradictions engendered by Soviet hegemony, and were made possible by the nationalism and determination of the Yugoslav and Albanian leaders.

The Helsinki Accords (1976) gave the official stamp to the de facto partition of the European continent.

Iron Curtain	▬
USSR	
Warsaw Pact	
Other Socialist countries	
Neutral states	
NATO nations	
Western allies	

The USSR has retained control of its shield in Eastern Europe, despite some violent crises (Berlin, 1953; Budapest, 1956; Czechoslovakia, 1968; Poland, 1980–1982).

Begun by the Marshall Plan, the recovery of Western Europe, which between 1945 and 1973 benefited from very favorable world economic conditions, was consolidated around the European Economic Community (the EEC), to which ten countries have gradually adhered. Despite various proposals, the EEC has led neither to political integration nor to the elaboration of a strictly European common defense within the Atlantic Alliance.

Western Europe, as it is defined in the framework of the EEC—and consequently the middle powers that belong to it—seems to have regional interests articulated around Eurafrica and the Mediterranean basin. West Germany, the leading economic power in Western Europe, remains politically determined by its past, the division of the German nation, and its geographical situation to the east.

For Germany as for France, the Franco-German alliance represents the heart of Europe. For many years France was bogged down in Indochina and Algeria in rearguard wars. After leaving NATO, while remaining a member of the Atlantic Alliance, France acquired nuclear firepower and laid down a strategy of self-protection based on deterrence of the strong by the weak. It is France that, politically and diplomatically, has conceptualized and continues to guarantee the Eurafrican—and Mediterranean— strategy for Europe, guided by a foreign policy laid down by General de Gaulle.

In spite of a distinct decline, the only other nuclear power, the United Kingdom—a latecomer into the Common Market—continues to retain a not insignificant weight in a number of sectors (finance, nuclear power, shipping, oil).

Great Britain, the creator of the first worldwide colonial empire, which laid the basis of the contemporary preeminence of the English language, has also been responsible for a large share of postcolonial conflicts and crises: the Indian subcontinent, the Israeli–Arab conflict, Cyprus, and so on. Great Britain has gradually abandoned most of the strategic positions it held, including the Persian Gulf region, to the benefit of the United States (1971). The Falklands war (1982) demonstrated the political determination of an old European power, its logistical capability, and the high quality of its professional army.

Italy, the fourth great European power, has vigorously industrialized over the last three decades. But the problem of the backwardness of its Mezzogiorno remains unchanged.

In the north, bordering neutral Finland and Sweden, NATO member Norway has a vulnerable Arctic position opposite one of the key zones of the Soviet military system.

The current debate concerning the Soviet SS-20s and the possible installation of American Pershings in the European theater so as to reestablish a credible deterrent touches directly on the balance on which the peace is based.

Gibraltar is the only interstate problem (United Kingdom–Spain) remaining in Europe.

Europe and the East

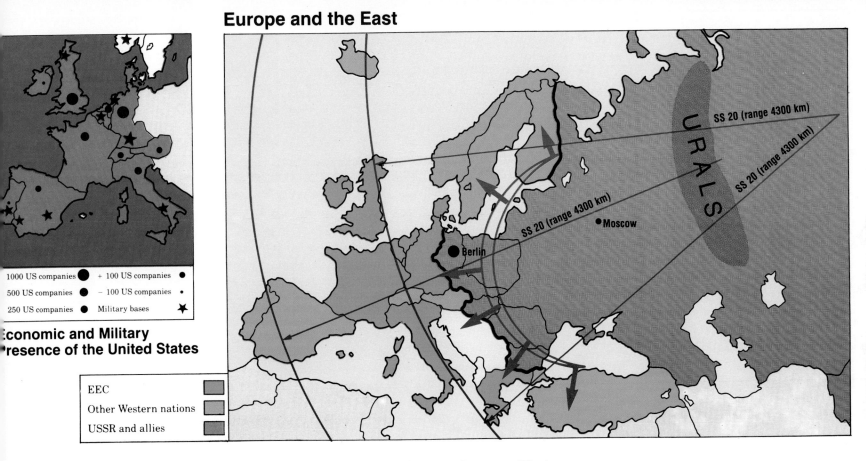

1000 US companies ● + 100 US companies ●
500 US companies ● – 100 US companies ·
250 US companies ● Military bases ★

Economic and Military Presence of the United States

EEC
Other Western nations
USSR and allies

URALS

SS 20 (range 4300 km)
SS 20 (range 4300 km)
SS 20 (range 4300 km)

● Moscow

● Berlin

Soviet military power is felt as a threat to Western
Europe.

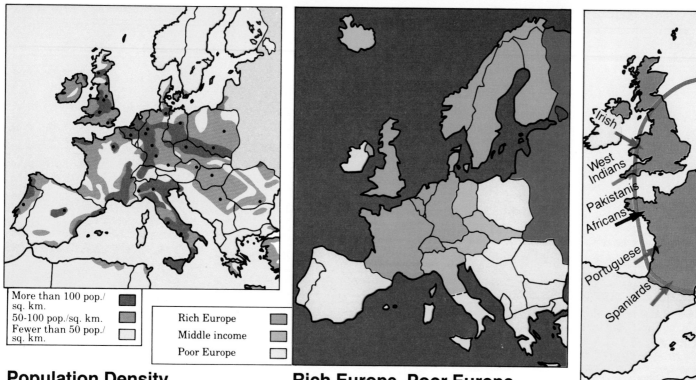

Population Density

Rich Europe, Poor Europe

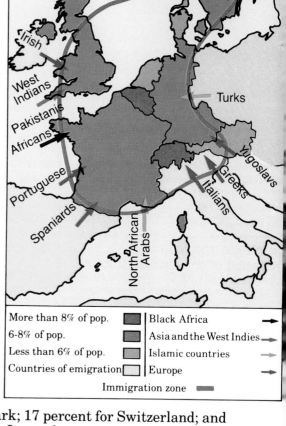

Immigrants in Western Europe

Western Europe is particularly heavily populated and, like all industrialized areas, has a very low population growth rate.

The EEC includes the most prosperous states (except for neutral countries such as Sweden and Switzerland, a financial center of the highest importance).

Europe, like the United States and the Persian Gulf countries, is a magnet for migrant workers (about 11 million). The percentage of migrants hovers around 7.5 for France, West Germany, the United Kingdom, and Belgium; 6 percent for Austria; 5 percent for Sweden; 4 percent for the Netherlands; 1.5 percent for Denmark; 17 percent for Switzerland; and over 30 percent for Luxembourg.

A source of hard currency for their home countries, the immigrant workers constitute a pool of cheap labor for the host countries; they pose a special political problem in times of economic crisis for the host countries and a source of dislocation in the event of massive repatriation to their homelands.

THE BASQUE PEOPLE

**LINGUISTIC
CONFLICT
IN BELGIUM**

Flemish

Walloon

Bilingual

German
speaking

German
Walloon

Flemish
Walloon

THE IRISH QUESTION

Catholic

10 to 30%   50 to 80%

30 to 50%   80% plus

Minority Problems

Caught between the idea of the nation-state and the idea of human rights, both born of the Enlightenment, and both of which mark the emergence of individual rights as well as modern nationalism, the rights of minorities (particularly ethnic ones) have been largely neglected. It is often a crucial problem in Third World countries, but minorities in Europe, including those in the liberal democratic countries, have suffered or still suffer more or less overt oppression or pressure for assimilation. In Central Europe: Hungarians in Transylvania (Romania); Albanians in Yugoslavia; and so on. In Western Europe: Irish Catholics in Ulster; Catalans and Basques in Spain; and, in a more specific way, the Flemish-Walloon conflict (originating in Walloon predominance until recently) in Belgium.

These more or less conflict-laden situations, often the result of pressures toward centralism, are generally resolved by either autonomy or federalism.

Unlike ethnic difficulties elsewhere, in Western Europe minorities are internal problems that do not precipitate destabilizing interference from other states.

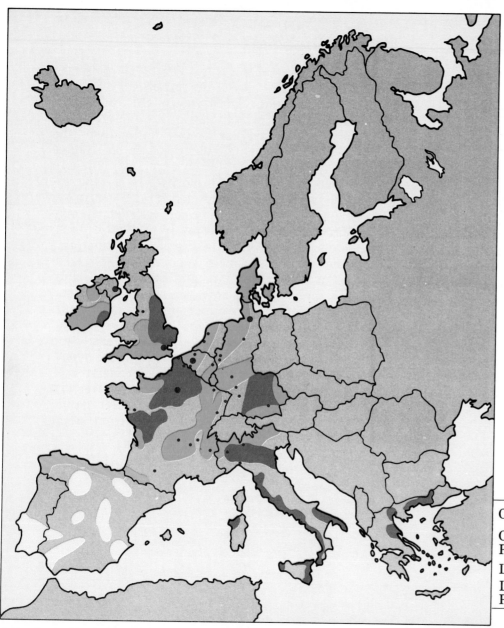

Main Agricultural Regions in the EEC

Unlike its European neighbors, France is one of the leading world exporters of grain. Europe's livestock industry is important. Agribusiness is developing rapidly and, in a world context, agriculture remains one of Europe's best assets.

Significance of the EEC in the World

(Percent of World Production)

Wheat	12%
Barley	22
Sugar	14
Milk	26
Meat	15
Corn	4

Note: Spain and Portugal have applied to join the EEC.

Cereals (EEC)	
Cereals (Spain, Portugal)	
Livestock (EEC)	
Livestock (Spain, Portugal)	

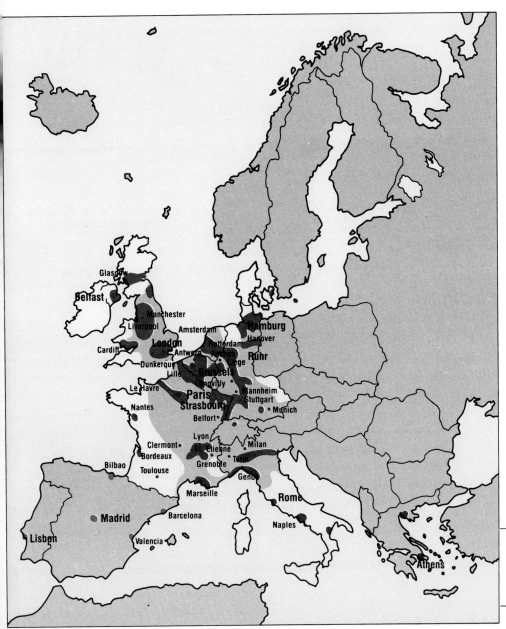

Europe: Industrial Power

The EEC is the world's second greatest industrial power—the leading one in terms of manufacturing output—just behind the United States.

Of the ten nations, West Germany alone produces almost half of most of the industrial goods produced by Europe.

Since the end of World War II, Europe's weakness in energy and mineral resources has continued to grow and constitutes—as it does for Japan, but to a lesser extent—a serious handicap.

EEC (% of world total)

Energy

Coal	11%
Hydroelectricity	16%
Nuclear power	20%
Oil	0.75%
Ores	slight

Output

Steel	18%
Aluminum	13%
Automobile manufacture	33%
Industrial equipment	13%
Shipbuilding	20%
Synthetic rubber	21%
Synthetic fibers	20%

Industrial areas

Principal centers

Underdeveloped areas

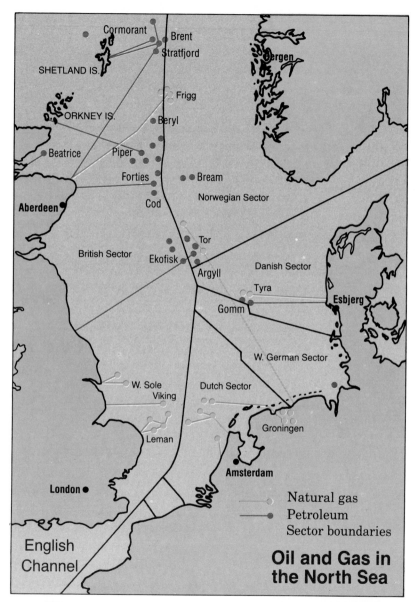

Oil and Gas in the North Sea

Cormorant
Brent
Stratfjord
SHETLAND IS.
Bergen
Frigg
ORKNEY IS.
Beryl
Beatrice
Piper
Forties
Bream
Cod
Norwegian Sector
Aberdeen
Tor
British Sector
Ekofisk
Argyll
Danish Sector
Tyra
Esbjerg
Gomm
W. German Sector
W. Sole
Viking
Dutch Sector
Groningen
Leman
Amsterdam
London
English Channel

○ Natural gas
● Petroleum
— Sector boundaries

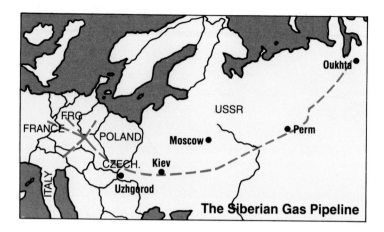

The Siberian Gas Pipeline

FRG
FRANCE
POLAND
CZECH.
ITALY
Uzhgorod
Kiev
Moscow
USSR
Perm
Oukhta

Western Europe and Its Energy Needs

To facilitate economic exchanges and to show its independence, the EEC has made a number of trade agreements:

- 1977—industrial free trade with the states of the former Free Trade Area (Switzerland, Austria, Sweden)
- 1978—periodic agreement with non-European states in the framework of GATT
- 1978—Lomé Convention, with a number of African, Caribbean, and Pacific and Indian Ocean states
- Various bilateral agreements (Yugoslavia, Romania)
- Agreements with Spain, Portugal, and Turkey, all candidates for admission to the EEC (Greece has been a member since 1980.)

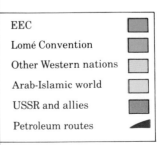

USSR and allies

Geopolitical and Geostrategic Aspects of Western Europe

The regional geostrategic area of Western Europe includes Africa, the Mediterranean basin, and the Persian Gulf, which determines policy toward the Arab states. By signing the Lomé Convention (1979) with 43 African states, the EEC created a system for regulating economic relations and thus made it possible to hope for relative stability in its traditional area of activity in Africa. Several small states in the Caribbean and the Pacific were also signatories to this convention.

The Organization of African Unity (OAU)
Founded 1963. Includes all the African states, as well as the Saharan Arab Democratic Republic and the islands of Cape Verde, São Tomé and Principe, Mauritius, the Comoros, and the Seychelles—except Namibia and South Africa. The problem of recognition of the Saharan Arab Democratic Republic, supported by Algeria and opposed by Morocco, divides the OAU.

Atlantic Ocean

Indian Ocean

EEC	
Lomé Convention	
Other Western nations	
Arab-Islamic world	
USSR and allies	
Petroleum routes	

MOROCCO · TUNISIA · SYRIA · IRAQ · ISRAEL · JORDAN · IRAN · AFGHANISTAN · ALGERIA · LIBYA · EGYPT · UNITED ARAB EMIRATES · SAUDI ARABIA · OMAN · CAPE VERDE · MAURITANIA · MALI · NIGER · CHAD · SUDAN · YEMEN · YEMEN · SENEGAL · GAMBIA · GUINEA BISSAU · GUINEA · UPPER VOLTA · BENIN · NIGERIA · DJIBOUTI · SIERRA LEONE · IVORY COAST · GHANA · TOGO · CAMEROON · CENTRAL AFRICAN REP. · ETHIOPIA · LIBERIA · S. TOMÉ · EQUAT. GUINEA · GABON · CONGO · UGANDA · KENYA · SOMALIA · RWANDA · BURUNDI · ZAIRE · TANZANIA · COMOROS · ANGOLA · MALAWI · ZAMBIA · MOZAMBIQUE · MADAGASCAR · MAURITIUS · RÉUNION · NAMIBIA · ZIMBABWE · BOTSWANA · LESOTHO · SOUTH AFRICA · SWAZILAND

Africa

Sub-Saharan Africa had remained largely outside East–West confrontations until 1975, but since the withdrawal of Portugal and the radicalization of Ethiopia (1977), it has ceased to be a Western preserve. Until that date, the major strategic issues in the Third World had been fought out along the peninsular belt that stretches from the eastern Mediterranean to the Far East. The USSR had made its presence felt in Egypt, Guinea, Zaire, Sudan, and Somalia, though generally without much success: withdrawal from Egypt, failure in Zaire, cool relations with Guinea, withdrawal from Sudan, expulsion from Somalia. The end of Portuguese colonialism created a vacuum that the USSR boldly exploited in concert with Cuba, and that created a new situation militarily, in both Angola and Ethiopia.

Although what is at stake in Africa is not vital, either for the United States or for the Soviet Union—although it is for Europe—southern Africa has an important role to play in terms of access to raw materials essential to the West and to Japan. South Africa is the only regional power there.

At the end of two decades of independence, the balance sheet in terms of development in sub-Saharan Africa is meager. The conditions typical of most African states are decline or stagnation of agriculture, and economic stagnation for the vast majority that lack a raw materials export base.

Contrary to the desires of the Organization of African Unity, all the inter-African associations launched over the last twenty years have collapsed. Even the principle of the inviolability of frontiers, which had been respected over quite a long period, has been broken over the last few years by sovereign states: the Somali-Ethiopian war over the Ogaden; annexation by Libya of a strip of land 100,000 kilometers square in northern Chad.

With an average rate of population increase of over 2.5 percent, projections give the following growth: 220 million in 1950, 350 million in 1970, and over 800 million in 2000.

To handicaps of recent origin, such as Balkanization, monoculture, economies geared to serve outside needs, and heavy cultural dependence, must be added others with deeper roots: ethnic divisions and antagonisms; low levels of social stratification, and low levels of productive forces.

Africa south of the Sahara is marked by weak states and virtually nonexistent nations.

It seems unlikely that there will be any major changes in the short term in the continuation of Western, and particularly European, influence in Africa. But unlike the Soviet assistance that turned sour in Egypt and Somalia, the Cuban military presence guarantees the survival of a Soviet presence insofar as it ensures the survival of existing regimes. This is particularly true in Angola.

With a Eurafrican geopolitical perspective, the EEC signed the important Lomé Convention (1979) and set up a stabilization system for export receipts for basic products, accepted the duty-free entry of most products exported by its African associated states, and substantially increased its annual assistance ($4 billion).

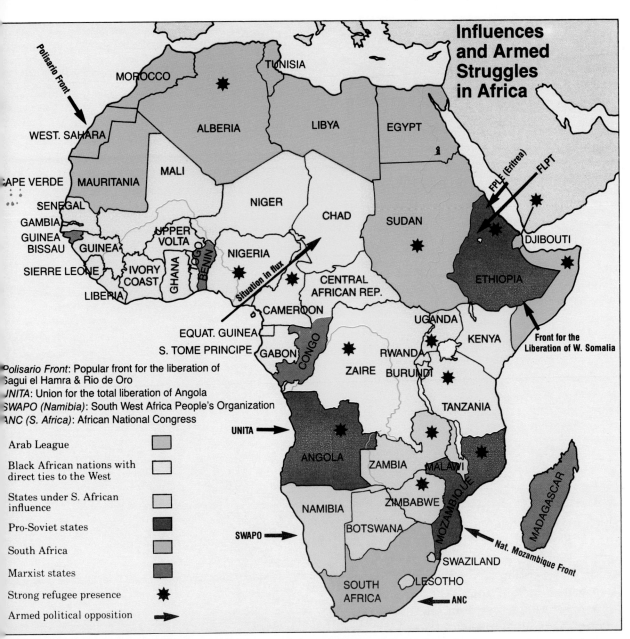

Influences and Armed Struggles in Africa

Polisario Front

WEST. SAHARA

MOROCCO
ALBERIA
TUNISIA
LIBYA
EGYPT
MALI
MAURITANIA
CAPE VERDE
SENEGAL
GAMBIA
GUINEA BISSAU
GUINEA
SIERRE LEONE
LIBERIA
IVORY COAST
GHANA
TOGO
BENIN
UPPER VOLTA
NIGER
NIGERIA
CHAD
SUDAN
CENTRAL AFRICAN REP.
CAMEROON
EQUAT. GUINEA
S. TOME PRINCIPE
GABON
CONGO
ZAIRE
UGANDA
RWANDA
BURUNDI
KENYA
TANZANIA
ANGOLA
ZAMBIA
MALAWI
ZIMBABWE
MOZAMBIQUE
NAMIBIA
BOTSWANA
SWAZILAND
LESOTHO
SOUTH AFRICA
MADAGASCAR
ETHIOPIA
DJIBOUTI

FPLE (Eritrea)
FLPT
Front for the Liberation of W. Somalia
Situation in flux
UNITA
SWAPO
Nat. Mozambique Front
ANC

Polisario Front: Popular front for the liberation of Sagui el Hamra & Rio de Oro
UNITA: Union for the total liberation of Angola
SWAPO (Namibia): South West Africa People's Organization
ANC (S. Africa): African National Congress

- Arab League
- Black African nations with direct ties to the West
- States under S. African influence
- Pro-Soviet states
- South Africa
- Marxist states
- Strong refugee presence ✳
- Armed political opposition →

Military Presence of the Soviets and Their Allies South of the Sahara

The only other African country where there are more than 1,500 military or civilian personnel from the Soviet bloc is the Congo. Between 1975 and 1982, the major beneficiaries of Soviet arms sales were Ethiopia, Angola, and Mozambique.

	USSR	Cuba	GDR*
Angola†	700	18,000	450
Ethiopia	2,400	5,900	550
Mozambique†	500	1,000	100

Source: Department of State, Washington, DC, 1982.

*The German Democratic Republic generally trains police and security personnel.
†UNITA and the Mozambique National Front are supported by South Africa.

FPLE: Popular Front for the Liberation of Eritrea
FLPT: Popular Front for the Liberation of Tiger.

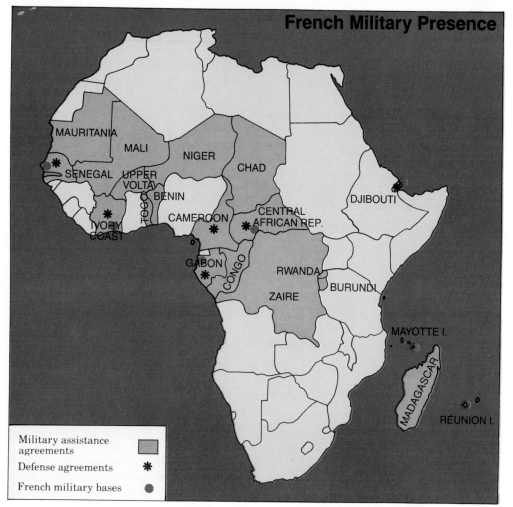

French Military Presence

MAURITANIA
MALI
NIGER
SENEGAL
UPPER VOLTA
TOGO
BENIN
IVORY COAST
CAMEROON
CHAD
CENTRAL AFRICAN REP.
DJIBOUTI
GABON
CONGO
RWANDA
ZAIRE
BURUNDI
MAYOTTE I.
MADAGASCAR
RÉUNION I.

Military assistance agreements
Defense agreements
French military bases

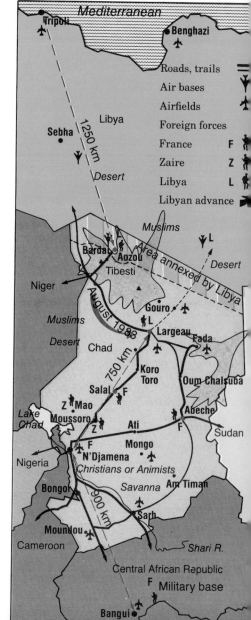

Chad

Mediterranean
Tripoli
Benghazi
Roads, trails
Air bases
Airfields
Foreign forces
France F
Zaire Z
Libya L
Libyan advance
Libya
Sebha
Desert
1250 km
Muslims
Barda
Aozou
Tibesti
L
Area annexed by Libya
Niger
Gouro
Desert
Muslims
L
August 1988
Largeau
Fada
Chad
Desert
Koro Toro
Oum Chalouba
750 km
Salal F
Z Mao
Abeche
Lake Chad
Moussoro
Z
Ati
Sudan
Nigeria
Mongo
N'Djamena
Christians or Animists
Am Timan
Savanna
Bongor
900 km
Moundou
Sarh
Cameroon
Shari R.
Central African Republic
F Military base
Bangui

The military initiatives of Libya, which had already annexed unilaterally the Aozou area (100,000 sq. km.), provoked in Chad operation Manta, led by French forces (August 1983). The latter had been withdrawn in 1980 after twelve years of intervention.

Conditions in Chad (deserts, lack of water, lack of population) pose delicate and costly logistical problems for the forces present there. A prolonged "peacekeeping" situation is unfavorable to France.

The Vulnerability of Africa
The Military Dimensions of the French Presence in Africa

France has defense agreements with several African countries—Djibouti, Gabon, Ivory Coast, Senegal, Cameroon, Central African Republic, and Togo—and military technical assistance agreements with these same countries, plus Mauritania, Niger, Upper Volta, Benin, Congo, Madagascar, Mali, Burundi, Rwanda, and Zaire.

Agreements on joint military maneuvers are in force with Djibouti, Gabon, Ivory Coast, Senegal, Togo, and Zaire. The total number of French military advisers in Africa is close to a thousand (the largest missions are in Djibouti, Gabon, Zaire, and the Ivory Coast). There are about 7,000 troops stationed on the continent, including 3,500 in Djibouti, 1,200 in Senegal, and 1,100 in the Central African Republic. French troops have intervened several times in Africa, notably: Gabon (1964), Chad (1968–1980), Mauritania (1977–1978, air support against the Polisario), Djibouti (1976–1977), Zaire (1977–1978), and Central African Republic (1979).

During the 1970s, France withdrew militarily (after a dozen years of intervention) from Chad (1980) and from Diego-Suarez in Madagascar (1975).

At the present time, the major strategic bases in Africa are Dakar, Gabon, Central African Republic, Djibouti, and Réunion Island.

France has a 25,000-man* rapid intervention force stationed on its territory, with logistical support that has undergone continuous improvement since 1978 (the Shaba operation in Zaire). It may be that this force is currently able to intervene more rapidly than that of other powers, including the USA and the USSR.

In twenty years, French positions in Africa have suffered little erosion. In various ways, France supplies annual aid totaling nearly a billion dollars. The policy of cooperation formulated by General de Gaulle has basically continued up to the present day. These alliances, based on political, economic, and cultural interests (most are with former colonies), have been shaped over several decades by the governing elites and France and seem likely to endure.

There are about 200,000 French technical assistants in "French-speaking" Africa. In these countries, as in other African countries, France is a major arms exporter.

As for religion, the importance of Islam in northern and central Africa is great and spreading. In the southern half, Christianity is widespread. There are also numerous animists.

The states of tropical Africa are marked by religious diversity, ethnic fragmentation, and weak political structures. Except for Nigeria and Ethiopia, the states are underpopulated. These special characteristics favor ethnically based strategies, attempts at destabilization, and interventions by troops from outside the continent. Tropical Africa remains highly vulnerable.

---

*A rapid action force of about 47,000 men is being set up. It is to be used in the European theater as well.

Angola

Railways —
S. African incursions ↑ ↑
Unstable areas ▒

The South African forces want to create a buffer zone extending about 200 km. inside Angola so as to make the guerrilla operations of SWAPO (the Namibian liberation movement), ineffective. Pretoria supports UNITA, which is fighting the regime in Luanda.

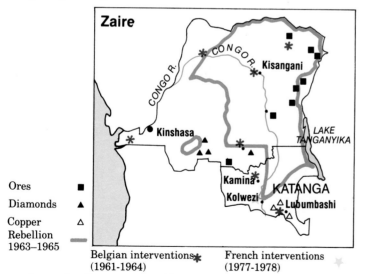

Zaire

Ores ■
Diamonds ▲
Copper △
Rebellion 1963–1965 ▬

Belgian interventions ✳ (1961-1964) French interventions (1977-1978)

Zaire was and remains the most unstable state on the continent. Between 1960 and 1965, then again in 1977 and 1978, it was the scene of outside interventions, as the West sought to protect its interests. Zaire continues to be the soft underbelly of Africa.

Southern Africa

Southern Africa is the richest part of the continent. The apartheid system makes the situation in South Africa more conflict-laden, especially since the disappearance of Portuguese colonialism and Soviet-Cuban involvement in Angola and, to a lesser extent, in Mozambique.

In the forefront of South African concerns is the Namibian shield, rich in minerals and underpopulated. South Africa possesses there the strategic enclave of Walvis Bay. The establishment of a regime favorable to Pretoria would allow it to keep the Soviet threat in check. It is to achieve this goal that the South African army is fighting SWAPO, the Namibian liberation movement backed by Angola. SWAPO is based on the Ovambo, an ethnic group that includes some 50 percent of the black population of Namibia (totaling slightly over 1 million). For its part, South Africa is supporting UNITA, an Angolan movement based on the largest ethnic group in the country and fighting the Luanda government.

Despite widespread hostility, South Africa is continuing its military operations, punctuated by deep incursions inside southern Angola. Talks between Angola and South Africa are unlikely to change a situation fraught with the prospect of prolonged conflict.

The relatively moderate position of Mozambique, which welcomes and trains militants of the African National Congress (ANC) hostile to Pretoria, has not led to direct military action by South Africa. But South Africa supplies and supports elements of the Mozambique National Front (MNF), which is hostile to the regime.

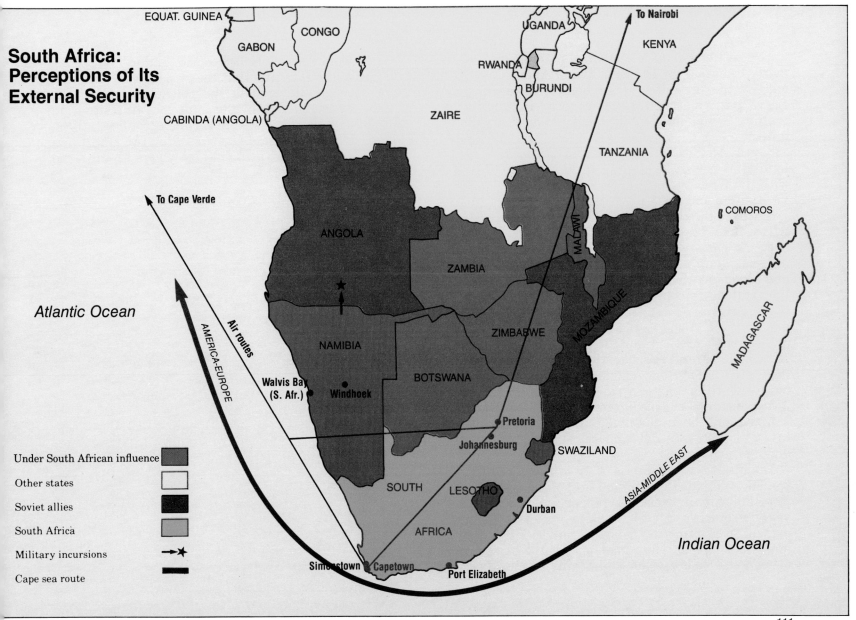

South Africa: Perceptions of Its External Security

Atlantic Ocean

Indian Ocean

To Cape Verde

To Nairobi

EQUAT. GUINEA

GABON

CONGO

UGANDA

KENYA

RWANDA

BURUNDI

CABINDA (ANGOLA)

ZAIRE

TANZANIA

COMOROS

ANGOLA

ZAMBIA

MALAWI

MOZAMBIQUE

MADAGASCAR

NAMIBIA

ZIMBABWE

BOTSWANA

Walvis Bay (S. Afr.)

Windhoek

Pretoria

Johannesburg

SWAZILAND

SOUTH

LESOTHO

Durban

AFRICA

Port Elizabeth

Simonstown

Capetown

Air routes

AMERICA-EUROPE

ASIA-MIDDLE EAST

Under South African influence

Other states

Soviet allies

South Africa

Military incursions

Cape sea route

111

The settlement of the conflict in Zimbabwe (formerly Rhodesia) by the British and Americans rules out for the present and for domestic reasons any active participation by that country in a regional conflict, especially as it is itself faced with serious unrest.

South Africa's strength lies in its military and industrial power, as well as in its de facto alliance with the West—given, among other things, its strategic position and its mineral resources.

Since 1973, thanks to NATO equipment, South Africa has possessed at Simonstown a detection network capable of reaching Australia and South America. South Africa's power feeds on the military and economic weakness of its adversaries. The front-line states (Tanzania, Botswana, Zambia, Mozambique, Angola) have no means of changing the status quo.

Despite theoretical hostility, trade between South Africa and several African states (Kenya, Zaire, Malawi, etc.) is far from insignificant. South Africa is, however, still a long way from organizing a "constellation of states" around Pretoria.

Apartheid, an unjustifiable system, gives advantages to Pretoria's adversaries and undermines its diplomatic position.

The numerical weakness of the white minority (15 percent) remains the Achilles' heel of the system, and this minority is declining proportionately year by year. Pretoria's racial strategy consists for the present in coopting the nonblack minorities (Indians, Malays, coloreds) while keeping out the blacks and deepending their divisions ethnically and juridically.

This policy seems to be too late and to involve too few changes. It is true that the principle of "one man, one vote" would lead to the political death of the white minority. But in the long run, this situation can only lead to increasingly violent confrontations. These will, of course, be exploited by the USSR and its allies in an African context where there is no support for apartheid.

South Africa

	Percent of World Production	Reserves
Gold	51%	Important reserves
Chromium	34	67%
Manganese	23	41
Vanadium	30	42
Platinum	45	81
Antimony	16	7
Ilmenite		15
Rutile		4
Asbestos	6	21
Diamonds	18	
Nickel	3	
Coal	4	
Uranium	14	11

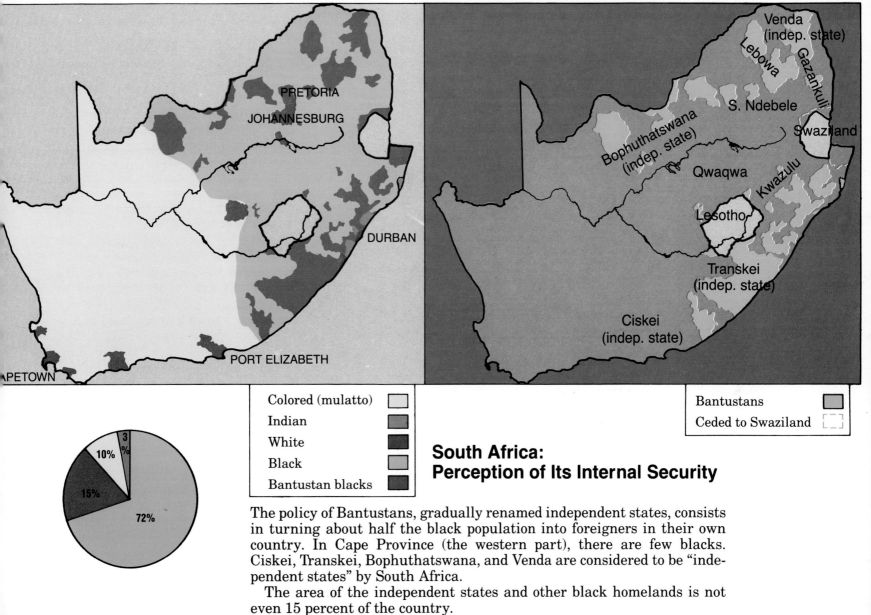

Colored (mulatto)
Indian
White
Black
Bantustan blacks

Bantustans
Ceded to Swaziland

PRETORIA
JOHANNESBURG
DURBAN
PORT ELIZABETH
CAPETOWN

Venda (indep. state)
Lebowa
Gazankuli
S. Ndebele
Swaziland
Bophuthatswana (indep. state)
Qwaqwa
Kwazulu
Lesotho
Transkei (indep. state)
Ciskei (indep. state)

3%
10%
15%
72%

South Africa: Perception of Its Internal Security

The policy of Bantustans, gradually renamed independent states, consists in turning about half the black population into foreigners in their own country. In Cape Province (the western part), there are few blacks. Ciskei, Transkei, Bophuthatswana, and Venda are considered to be "independent states" by South Africa.

The area of the independent states and other black homelands is not even 15 percent of the country.

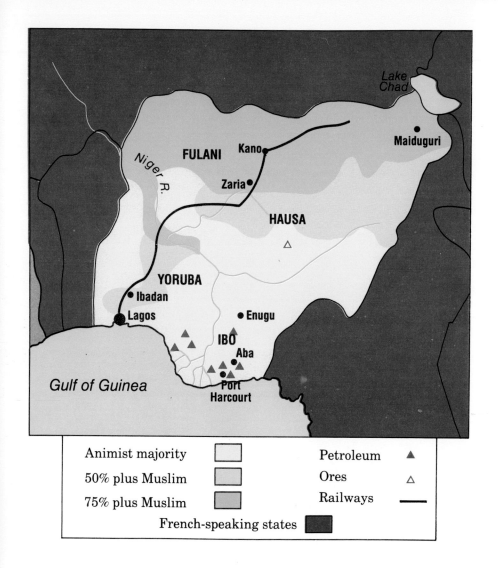

Animist majority

50% plus Muslim

75% plus Muslim

French-speaking states

Petroleum ▲

Ores △

Railways ——

Lake Chad

Maiduguri

FULANI **Kano**

Niger R.

Zaria

HAUSA

△

YORUBA

Ibadan

Lagos

Enugu

IBO

Aba

Gulf of Guinea

Port Harcourt

Nigeria: A Regional Power in Tropical Africa

Nigeria is the only state that can aspire to be a regional power in tropical Africa, and it is also the most populous country in Africa (about 90 million inhabitants).

The republic's federal system seeks to resolve the ethnic and religious problems that were largely responsible for the attempted Biafran secession (1967–1970).

With its large oil resources, Nigeria saw real growth in the 1970s, which enabled it to play a continent-wide political role. The direct impact of Nigeria on its neighbors is partly hampered by the fact that they are French-speaking. The relatively rapid end of the oil boom has brought Nigeria, like all the populous oil states, back to its former difficulties, as was shown by the expulsion of 2 million foreign immigrant workers to neighboring countries in 1983.

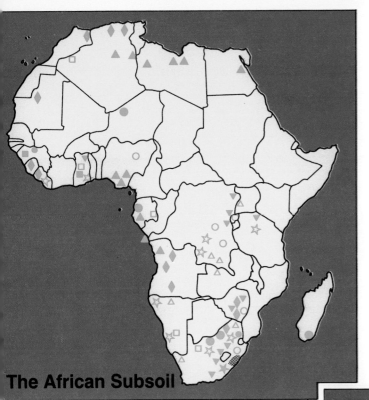

The African Subsoil

Petroleum	▲	Bauxite	▪
Uranium	●	Manganese	□
Copper	△	Tin	○
Iron	◆	Platinum	★
Chrome	●	Diamonds	☆
Gold	▼	Coal	▨

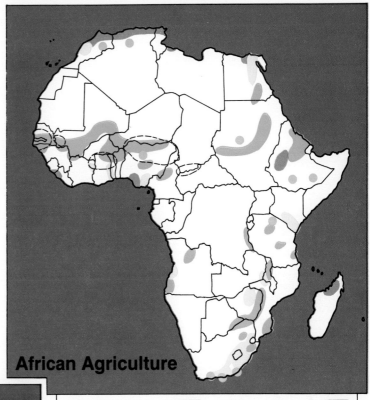

African Agriculture

Barley	▨	Rice	▨
Sheep	●	Wheat, corn	▨
Peanuts	⬭	Millet, sorghum	▨

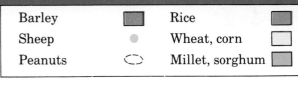

Sahel

LDCs □

The Less Developed Countries

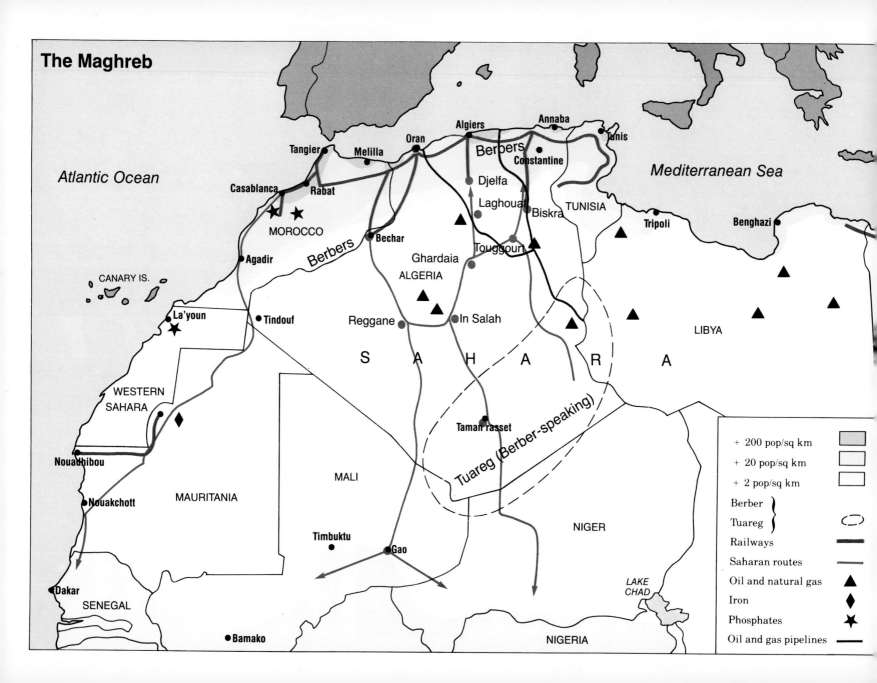

The Maghreb

Atlantic Ocean

Mediterranean Sea

Tangier
Melilla
Oran
Algiers
Annaba
Tunis
Berbers
Constantine
Casablanca
Rabat
Djelfa
Laghouat
Biskra
Tripoli
Benghazi
MOROCCO
TUNISIA
Berbers
Bechar
Ghardaia
Touggourt
Agadir
ALGERIA
CANARY IS.
La'youn
Tindouf
Reggane
In Salah
LIBYA
WESTERN
SAHARA
S A H A R A
Nouadhibou
Tamanrasset
Tuareg (Berber-speaking)
Nouakchott
MALI
MAURITANIA
NIGER
Timbuktu
Gao
LAKE
CHAD
Dakar
SENEGAL
Bamako
NIGERIA

+ 200 pop/sq km	
+ 20 pop/sq km	
+ 2 pop/sq km	
Berber 〉 Tuareg 〉	
Railways	
Saharan routes	
Oil and natural gas	▲
Iron	◆
Phosphates	★
Oil and gas pipelines	

The Maghreb extends, geographically, from Mauritania to Libya. It is peopled by Arab Berbers and has more than 50 million inhabitants, most of whom are concentrated in the coastal strip or on the slopes of the Atlas Mountains. After southern Africa, it is the richest region in Africa (oil and gas, phosphates, iron).

Maghreb unity, although an oft-proclaimed goal, remains in the realm of the hypothetical, and has been dominated for the last two decades by the rivalry between Morocco and Algeria, further fueled by the conflict in and over Western Sahara. One of the issues at stake in this conflict is dominance in the Maghreb. Morocco would win a virtual monopoly on the world supply of phosphates (outside the USA and the USSR); Algeria would obtain indirect access to the Atlantic and at the same time finally destroy Moroccan dreams of a "Greater Morocco."

Western Sahara (256,000 sq. km.) has the distinction of being the only decolonized African territory (in addition to Eritrea) not to have gained independence automatically. The Polisario Front, set up in 1973, first fought Spanish colonialism. In 1975, basing itself on what it asserted were its historical rights, Morocco occupied two-thirds of the then Spanish Sahara, from which Madrid had just withdrawn. Mauritania occupied a third of the country (1975–1979), and then gave it up. Heavily supported by Algeria, the Polisario carried on an effective guerrilla war (1976–1982), and won recognition, thanks to Algeria, from a large number of members of the OAU as the representative of the Saharan Arab Democratic Republic. Neither the USA nor the USSR recognizes the SADR. France supports the idea of a referendum.

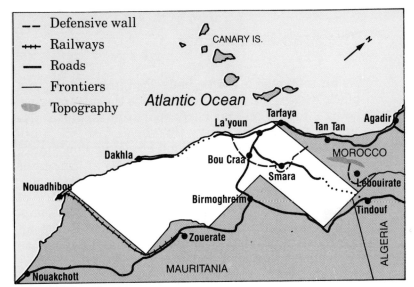

Western Sahara

The erection of a wall (1979) marking off the useful part of the country has turned out to be militarily effective.

Given the prevalence of nomadism and the number of refugees outside its borders, such a referendum would be likely to be disputed. The last Spanish census gave 73,500 inhabitants. A realistic estimate would put the figure between 100,000 and 150,000.*

Libya, a rich oil state that is very underpopulated, lacks the manpower to fulfill the ambitions of Colonel Qadafi. It has confined itself to the role of troublemaker, both in the Arab world and along the edges of the Sahara. The USSR is active there. Apart from Libya, all the Maghreb states send a substantial proportion of their workers to European countries.

---

*The Polisario Front gives a figure of 300,000 Sahrawis.

The Arab Middle East

As an intermediate zone between Europe, Asia, and black Africa, the Arab world enjoys the advantage of being situated on major transport routes, but it also suffers the vulnerability this gives rise to. Since World War II the strategic and economic importance of this vital area has been continuously at the center of world tensions.

The creation of Israel (1948), a consequence of modern nationalism, anti-Semitism, and the genocide of the Jews, led to Arab rejection of European interference. From 1967, Palestinian nationalism has made its presence felt and raises not only the refugee problem, but also that of the necessity for a separate state.

The Arab world, confronted by an industrialized Israeli nation that possesses a modern army and is supported by the USA, has been mobilized and inspired around slogans of unity and demands for a Palestinian state.

The Arab countries, whose political autonomy was limited until 1952 and even until 1958, progressively built up their strength until, from 1973 onward, they became an economic and financial factor in the world balance of power. Nationalism, which used to be predominantly socialistic (Nasserism and Baathism), has today become conservative, partly because of the place occupied by Saudi Arabia. Militarily, the neutralization of Egypt (1978) gave a decisive boost to Israeli superiority. The PLO lacks both the strength to impose its demands and the means to negotiate (negotiation being seen as treason).

Israel, obsessed with security and enjoying military superiority, has annexed the Golan Heights (which belonged to Syria) and, by multiplying its settlements, demonstrated its determination to retain the West Bank (Judea and Samaria). The number of Palestinians under Israeli jurisdiction is about 2 million (of whom 1.4 million are in the occupied territories), as against 3 million Jewish Israelis.

The Arab world has vast mineral resources: more than 50 percent of world oil reserves and 25 percent of gas reserves, a third of the world's phosphate, and an abundance of several vital minerals. The agricultural sector remains weak. Above all, too few concerted efforts seem to have been made to take maximum advantage of the financial resources at its disposal to lay the foundations for real agricultural and industrial development.

Islam, "religion and way of life," proclaims itself to be a social and cultural fact defining every aspect of behavior. It is interpreted today not only as a religion, but also as a national value. Only rarely is it not the state religion. National values have often been given pride of place when modern Arab nationalism has asserted itself. In periods of crisis like today, fundamentalism exercises a growing influence.

Religious sects continue to evoke strong feelings of belonging. In Syria, for example, state power is tightly controlled by the Alawite minority, in Iraq by the Arab Sunni minority.

Baathist Syria, rival of the no less Baathist Iraq, an ally of the USSR, is isolated. It has failed to bring to fruition its plans for a Greater Syria, including dragging Lebanon and Jordan into its orbit. The essential problem of the government there appears to be self-preservation.

Jordan has steadily strengthened itself over the last decade both militarily and economically, thanks largely to American aid. Although often threatened, it has succeeded in surviving, thanks to the political skill and determination of King Hussein. The Hashemite kingdom is in a peculiar situation in that two-thirds of its population is Palestinian. A federal Jordanian–Palestinian arrangement has often been proposed. Among other things, it would involve Israeli withdrawal from the West Bank, and that seems most unlikely.

The region's strategic importance involves Great Power support from the nations' respective allies. Despite the efforts of the USSR, the USA has managed to retain its predominance in the region. Its problem remains to reconcile, insofar as that is possible, the divergent aspirations regarding the Palestinian problem and the status of Jerusalem of its Israeli ally, on the one hand, and on the other, of its Arab allies, especially Saudi Arabia.

Given that a settlement of the Palestinian problem seems unlikely, the prospects of continuing conflict in the Middle East remain great.

The League of Arab Nations: Created in 1945, it comprises all 21 Arab states except Western Sahara (formerly Spanish). It includes one non-Arab state, Somalia. The PLO (Palestinian Liberation Organization) is, in fact, an equal member. Many economic associations reinforce cooperation.

Islam and Religious Sects in the Middle East

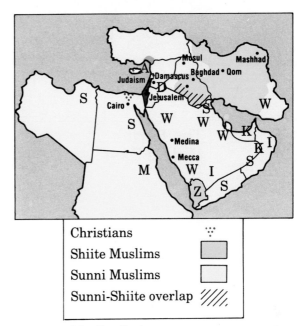

Christians	⋮
Shiite Muslims	▨
Sunni Muslims	☐
Sunni-Shiite overlap	▨

Muslim Sects

Wahhabites (W)

Alawites (A)

Druse (D)

Zaidites (Z)

Sanoussis (S)

Mahadis (M)

Ismailis (I)

Kharijites (K)

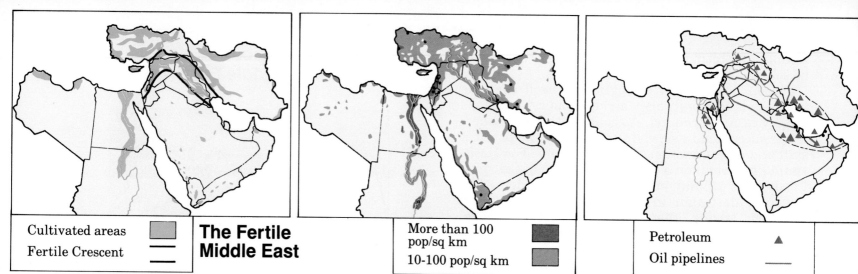

The Fertile Middle East

Cultivated areas

Fertile Crescent

Sparse Populations

More than 100 pop/sq km

10-100 pop/sq km

A Major Resource: Oil

Petroleum

Oil pipelines

Land and Water	% Land Under Cultivation	% Irrigated Cultivated Land
Iran	14	20
Turkey	34	7
Israel	20	40
Saudi Arabia	1	80
Egypt	2.5	100
Iraq	18	50
Libya	2	9
Syria	35	10
Lebanon	27	20
Jordan	11	7
Yemen	7	10
South Yemen	1	80
Kuwait	1	100
United Arab Emirates	5	100
Oman	1	—
Mean	11	24

Oil in the Gulf*

Production 1981		Reserves (54% of world total)
Saudi Arabia	63.6%	46%
Emirates	9.7%	9%
Iran	8.5%	16%
Kuwait	7.5%	15%
Iraq	5.8%	9%
Qatar	2.5%	1%
Oman	2.0%	0.5%
Bahrain	0.4%	0.1%

*The Gulf contains 25% of the world reserves of natural gas.

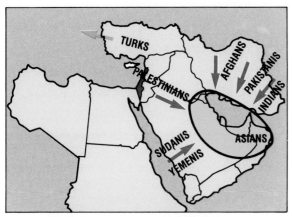

Immigration into the Gulf

Thanks to various means of fuel conservation and substitution, the West has succeeded, in ten years, in reducing its oil consumption by about 25 percent.

Oil production had to be progressively reduced after 1979–1980. In 1983, oil prices fell. Over the decade 1973 to 1983, the percentage of oil coming from the Gulf was 30 percent for the USA, 60 percent for Western Europe, and 70 percent for Japan.

Population of the Arabian Peninsula (in millions)

Saudi Arabia	10.4
Yemen	5.8
South Yemen	2.0
Kuwait	1.4
Oman	0.9
Emirates	0.8
Bahrain	0.4
Qatar	0.2
Total	**21.9**

Foreign Skilled Personnel and Labor in the Arab Gulf States

	Total Employed	Foreign Employed	%
Saudi Arabia	2,384,000	1,300,000	57
Bahrain	60,000	30,000	50
Emirates	300,000	239,000	80
Kuwait	304,000	211,000	75
Oman	350,000	70,000	20
Qatar	86,000	54,000	80

Origins:

Yemenis, Palestinians, Egyptians, Jordanians	75%	Pakistanis, Indians, Other Asian countries	25%

The Arabian Peninsula is the most vulnerable region in the Gulf. Five states combined (including those grouped in the United Arab Emirates) do not have more than 4 million inhabitants. It was essentially to guarantee the stability of Saudi Arabia and these five states that the American Rapid Deployment Force was set up. Part of this force can be deployed from Diego Garcia (3,700 km. from the Gulf), with facilities at the air and naval bases at Masirah (Oman), Berbera (Somalia), and Mombasa (Kenya). The main drawback of the RDF at present lies in the excessively long time required to intervene (between 3 and 7 days for the first forces). The very idea of this force seems too ponderous and powerful, for lightning operations are infinitely more likely than an intervention of conventional dimensions.

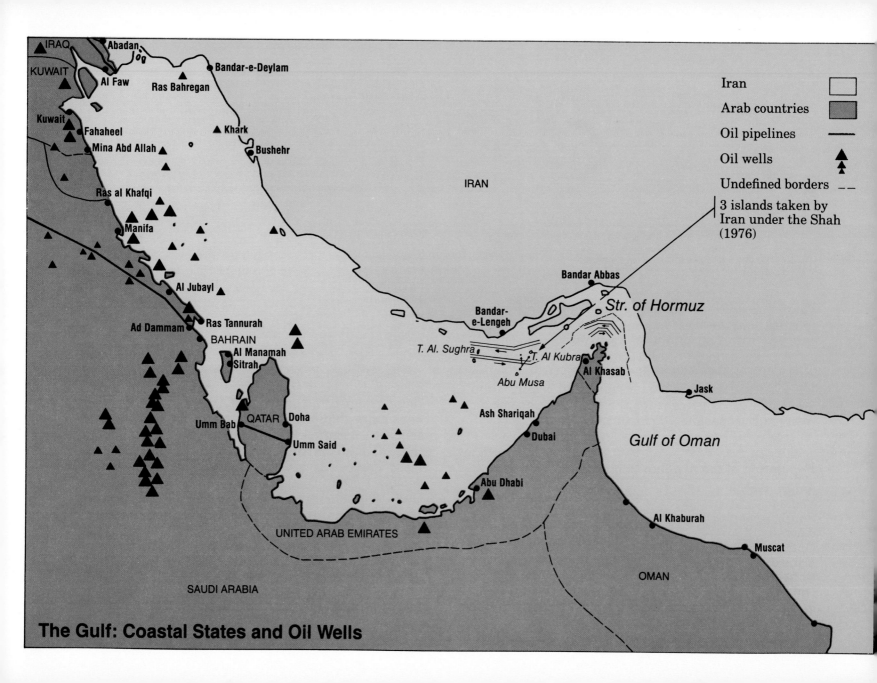

The Gulf: Coastal States and Oil Wells

Saudi Arabia

In November 1973, Saudi Arabia reduced its oil output by 30 percent and soon after quadrupled the price. In a year, its GNP rose 250 percent. Since 1974, Saudi Arabia's position, which had already been rising after 1967, has become preeminent. During the previous decade, the Wahhabite kingdom, founded on fundamentalist Islam and enduring tribal structures, had been mainly preoccupied in foiling attempted revolutions deriving from Middle East socialism and Nasserism in particular. Egypt's defeat in 1967, then Sadat's rise to power and the break with the USSR (1972), which it encouraged, enabled Saudi Arabia to go on the offensive as leader of the conservative forces in the Arab Middle East.

Although it has increased its capabilities, particularly its air power (air base at Tabuk, not far from Israel), Saudi Arabia remains a second-rank military power. The leaders of a kingdom that is underpopulated (about 10 million inhabitants), fabulously rich, and sharply stratified, in the middle of an unstable and coveted region, are obsessed with security and stability. Political control is highly centralized and tight, reinforced by the peculiar harshness of Wahhabism. The royal family (which is very large) incarnates the regime, and fears both any significant change in the world market that might precipitate a depression, and the effects of rapid modernization in a society built on traditional religious and tribal values. The occupation of the Great Mosque in Mecca in 1979 was a sign of the relative fragility of the regime. The proportion of immigrant labor is high, although it is split into various ethnic groups and is frequently rotated.

Externally, the Soviet presence in the Red Sea and the pro-Soviet revolutionary center in South Yemen are perceived as serious threats. The southern frontier is strategically vulnerable.

Saudi Arabia is too weak militarily to influence its neighbors, but uses its vast financial resources as a means of foreign policy. Its aid, very large in volume, is used to ensure the stability of friendly neighboring Arab regimes, to help the Arab states generally (in their struggle with Israel among other things), to combat vigorously any Communist influence, and to encourage Muslim states throughout the world (including those of Africa and Southeast Asia) to strengthen the place of Islam in state and society. More ambitiously, Saudi Arabia is attempting as much as possible to maintain the stability of the world market, hence its moderating role in OPEC. In this respect, alliance with the USA is seen as vital, and the interests of the two states converge (except over Israel). In a period when dangers are increasing everywhere, internal, regional, and world stability are at the center of Saudi preoccupations.

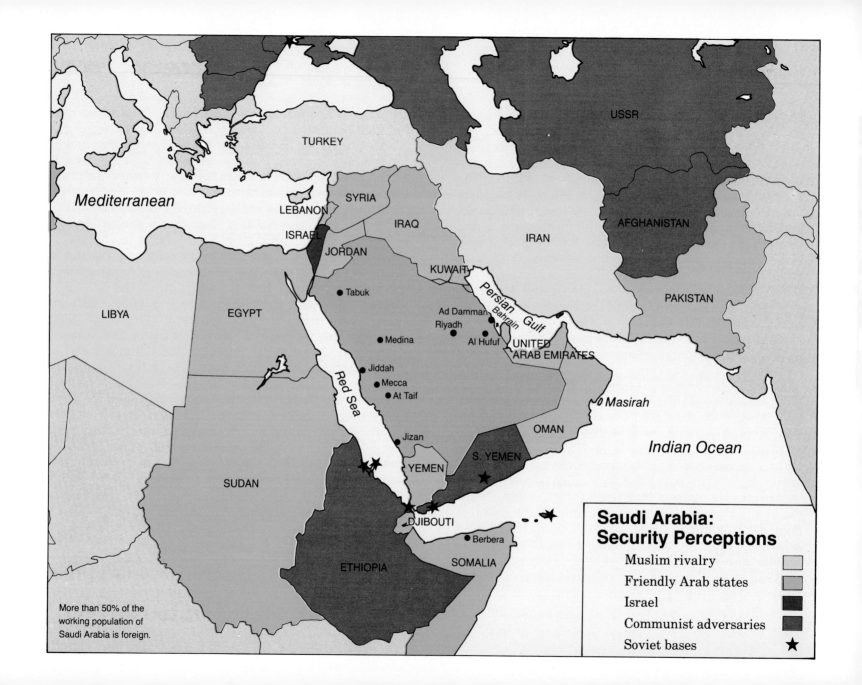

**Saudi Arabia:
Security Perceptions**

Muslim rivalry

Friendly Arab states

Israel

Communist adversaries

Soviet bases ★

More than 50% of the
working population of
Saudi Arabia is foreign.

Mediterranean

TURKEY

USSR

SYRIA

LEBANON

ISRAEL

JORDAN

IRAQ

IRAN

AFGHANISTAN

KUWAIT

PAKISTAN

LIBYA

EGYPT

Tabuk

Ad Damman

Bahrain

Persian Gulf

Riyadh

Medina

Al Hufuf

UNITED
ARAB EMIRATES

Red Sea

Jiddah

Mecca

At Taif

Masirah

Jizan

OMAN

Indian Ocean

S. YEMEN

YEMEN

SUDAN

DJIBOUTI

Berbera

ETHIOPIA

SOMALIA

EGYPT UNDER NASSER

CYPRUS
SYRIA
LEBANON
USA 1958 →
Suez 1956 ↓
JORDAN
ISRAEL
IRAQ
Cairo •
KUWAIT
LIBYA
EGYPT
SAUDI ARABIA
SUDAN
YEMEN
ETHIOPIA
ADEN

Members UAR
Other Arab states
Suez conflict →
Lebanon (US interv.) →

COLONIAL EGYPT

CYPRUS
SYRIA
LEBANON
PALESTINE
TRANS-JORDAN
IRAQ
EGYPT
SAUDI ARABIA
SUDAN
ANGLO-EGYPTIAN
BRITISH PROTECTORATE
ERITREA
ADEN
FRENCH SOMALIA
SOMALIA
ETHIOPIA

Under French control
Under British control
Under Italian control
British mandate

EGYPT'S GEOPOLITICAL LOGIC

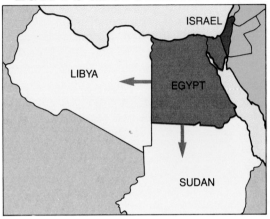

ISRAEL
LIBYA
EGYPT
SUDAN

Egypt

Egypt is the center of the Arab world: It is the most populous Arab state—one third of all Arabs are Egyptians—and it is also the most ancient and most homogeneous nation in the Arab world, with the largest agricultural population in the Middle East. In spite of very limited resources, Egypt remains, because of its population, a key factor in the Middle Eastern balance.

In Nasser's time, attempts at Arab unity brought Egypt and Syria together in the United Arab Republic (1958–1961), which in 1963 added Iraq and Yemen. These alliances were short-lived.

Considering its regional geopolitical interests rather than the utopia of pan-Arabism, Egypt should logically direct its efforts in the direction of Sudan (with its fertile, underpopulated lands) and Libya, rich in oil and very sparsely populated.

Today, since the conclusion of the peace treaty (1978) that enabled Egypt to regain Sinai (1982), Cairo has taken up a neutral position in possible future Arab–Israeli conflicts. The American alliance and Saudi aid seem unlikely to resolve the crisis of Egyptian society (in part a demographic one), any more than Nasserite "socialism" did before them. Geopolitical ambitions could in future appear as a solution, if the regional situation allows it.

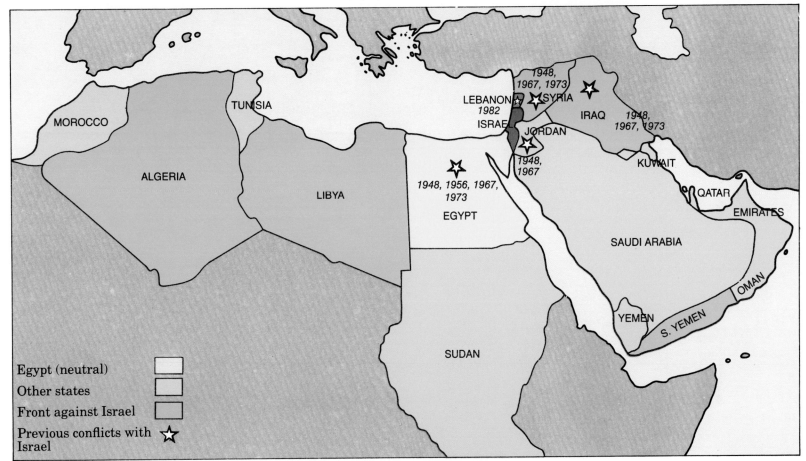

Israel and the Arab World

The agreement among some Arab countries never to recognize Israel, set up in December 1977, includes: Algeria, Libya, Syria, South Yemen, Iraq, and the PLO.*

According to UNRRA, there are 1,925,762 Palestinian refugees (1982). The Palestinian diaspora is estimated to number 2 million:

Jordan	920,000
Lebanon	420,000
Syria	250,000
Kuwait	200,000
Others	250,000

*Iraq's participation in the 1967 and 1973 wars was limited.

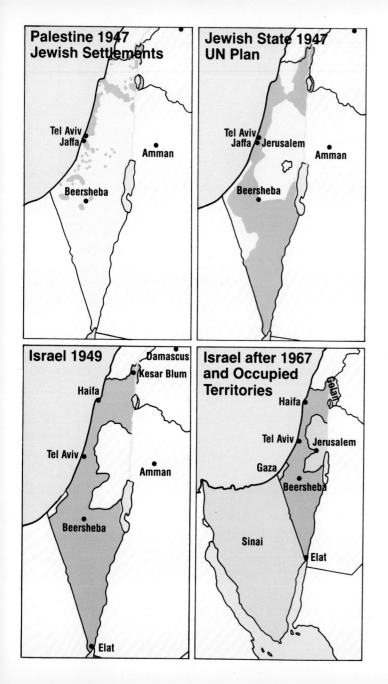

Palestine 1947
Jewish Settlements

Tel Aviv
Jaffa

Amman

Beersheba

Jewish State 1947
UN Plan

Tel Aviv
Jaffa
Jerusalem

Amman

Beersheba

Israel 1949

Damascus
Kesar Blum

Haifa

Tel Aviv

Amman

Beersheba

Elat

Israel after 1967
and Occupied
Territories

Haifa

Golan

Tel Aviv
Jerusalem

Gaza

Beersheba

Sinai

Elat

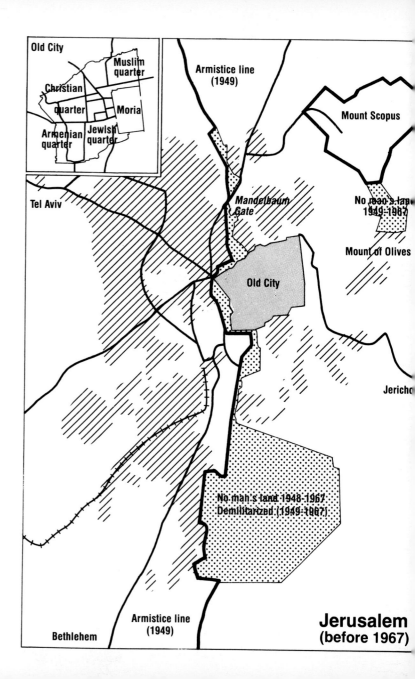

Old City

Muslim quarter

Christian quarter

Moria

Armenian quarter

Jewish quarter

Armistice line (1949)

Mount Scopus

Tel Aviv

Mandelbaum Gate

No man's land 1949-1967

Mount of Olives

Old City

Jericho

No man's land 1948-1967
Demilitarized (1949-1967)

Bethlehem

Armistice line (1949)

Jerusalem
(before 1967)

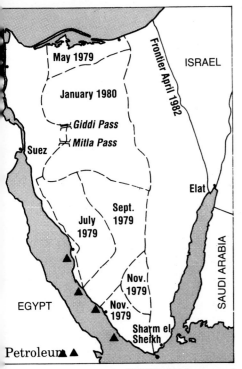

**Israeli Withdrawal
from Sinai
(1978–1982)**

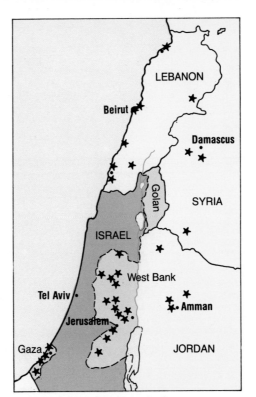

**Principal Palestinian
Refugee Camps ★**

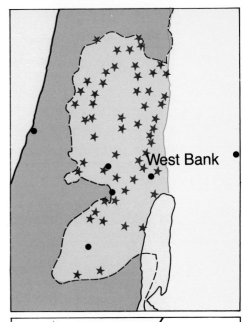

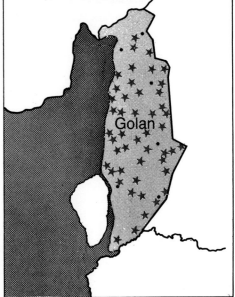

**Israeli Settlements:
West Bank and
Golan Heights**
★

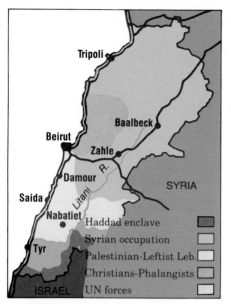

Lebanon, early 1982

Legend:
- Haddad enclave
- Syrian occupation
- Palestinian-Leftist Leb.
- Christians-Phalangists
- UN forces

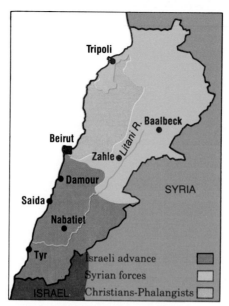

Lebanon, June 1982

Legend:
- Israeli advance
- Syrian forces
- Christians-Phalangists

**Lebanon, July 1983,
after the Israeli Retreat**

Legend:
- Syrian forces
- Israeli forces
- Israeli retreat
- Lebanese zone

After the removal of Palestinian organizations from Jordan (September 1970), most Palestinian forces regrouped in Lebanon, militarily the weakest state in the region. Their presence broke the fragile Lebanese communal and political balance. Civil war broke out in 1975 and allowed Syria to intervene (1976), fatally weakening Lebanese sovereignty, which rapidly disintegrated into various spheres of influence. Internal confrontations resulted in de facto partitions.

Israeli intervention (1982) as far as Beirut altered the relations of force inside the country by mandating the withdrawal of PLO forces. The sovereignty of the Lebanese state for the time being is purely nominal, and antagonisms are still very much alive. The laborious negotiations on the withdrawal of Israeli troops from Lebanon enables the much more important problem for Israel—the status of the West Bank and the future of its settlements—to be postponed.

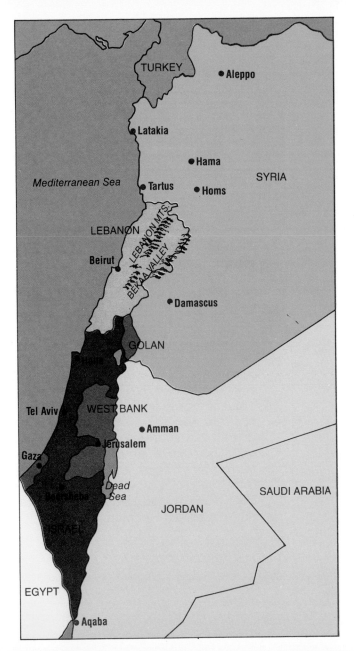

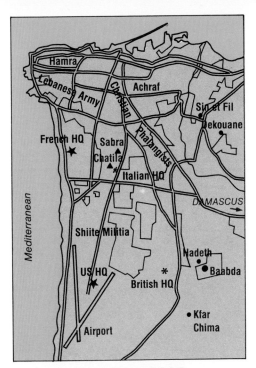

Beirut, October 1983

Lebanon and Its Environment

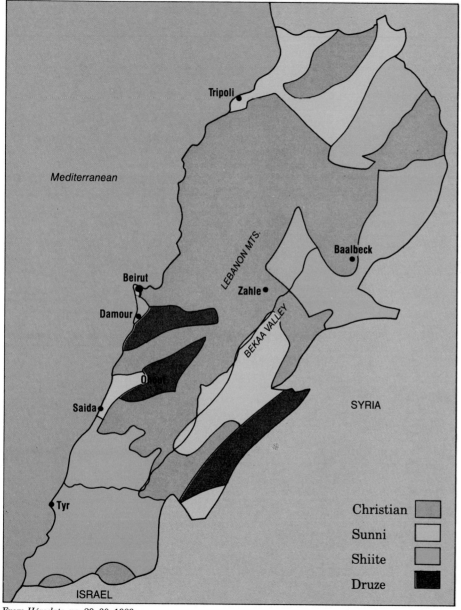

From *Hérodote*, no. 29–30, 1983.

Principal Religious Communities of Lebanon

Shiite	about 600,000 to 650,000
Maronite	about 550,000 to 600,000
Sunni	about 500,000
Greek Orthodox	about 250,000 to 280,000
Greek Catholic	about 200,000 to 220,000
Druze	about 150,000 to 180,000

- The figures given here for communities active as such in actual conflicts are estimates where there are no recent census data for these religious sects.
- The Shiites are spread out over a good part of the country in scattered settlements. Most are rural people.
- The Sunni are traditionally city people (Beirut, Tripoli, Saida). They occupy rural areas near the cities, as well as the Bekaa Valley.
- The Greek Orthodox and Greek Catholic communities are made up of Arabs Christianized over the centuries. They are traditionally city people, especially the Greek Orthodox group.
- The mountains of Lebanon and other hilly areas have historically been places of refuge for two threatened communities: the Maronite Christians and the Druze (a dissident sect of Islam). The Maronite villages are located on the exit routes from the cities as well as on the mountain routes. The Druze, in addition to the mountains, occupy the Shouf and Mt. Hermon.

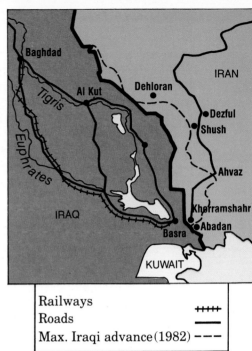

The Iran-Iraq Conflict

Railways ++++
Roads ———
Max. Iraqi advance (1982) ———

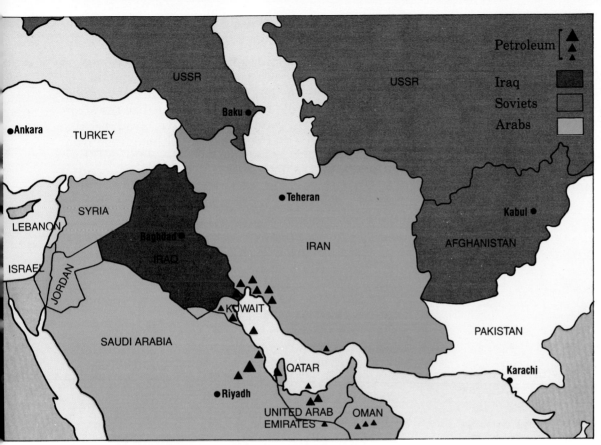

Iran's Perception of External Threats

The fall of the Shah after a popular revolution (1979) deprived the United States of a major ally in the region. None of the forces opposing the government of Khomeini and the mullahs has been able to gain widespread support, and the masses of the population continue to support a regime that seems destined to last, no doubt even after the disappearance of its charismatic leader.

But hostility toward the United States should not obscure the fact that it is with the Soviet Union that Iran has a long common frontier, and that its sovereignty was for a long time a matter of dispute between Russia and Great Britain, which even drew up a proposal for partition (1907). Moreover, by moving into Afghanistan, the USSR now adjoins Iran's eastern frontier.

As early as 1979, the Kurds (Sunni Muslims) demanded autonomy within the framework of Iran; they established this autonomy de facto and have maintained it since by force of arms. Most Iranians, welded together by Shiism and by Persian culture, both expressions of their identity, and fired by revolutionary ardor and/or the climate of the "fatherland in danger," offered fierce resistance to the Iraqi invasion (1980), helped by their heavy numerical superiority (3 to 1).

Iran–Iraq Conflict

The rivalry between Iraq and Iran (which for a long time supported Iraqi Kurds fighting the central government) was fueled by concessions Iraq was obliged to make to Iran in 1975 (the Algiers agreements) leading to withdrawal of Iran's support for the Kurds, and their eventual collapse. These concessions involved problems of sovereignty in the Shatt-al-Arab and the Iraqi claim to the ethnically Arab Iranian province of Khuzistan.

Capitalizing on the restoration of peace and the considerable increase in oil revenues, Saddam Hussein, the Iraqi strong man in a regime which, since 1968, had devoted itself to building and consolidating the state, launched an ambitious development program in Iraq.

Hussein took advantage of the disorder created by the revolutionary process in Iran and sought, by invading Iran, to make Iraq the leading regional power in the Middle East. Failing to utilize the strategy of the rapid tank-led offensive, Iraq became bogged down in a war of attrition. Although the Iraqi people were not very enthusiastic about this military adventure, the majority, who are Shiite, ignored the calls to revolt issued by Ayatollah Khomeini.

After two years, and despite great logistical and supply difficulties, the Iranian forces succeeded in turning the situation around. But, having invaded Iraq, the Iranians in turn became bogged down. The Baath party's organization (some 400,000 members) appears likely to hold up, while external assistance (from France* and the Arab countries, including Saudi Arabia) is assured.

No state (except Syria and Israel) wants the collapse of the Iraqi regime and a change in the balance in Iran's favor. Whatever the outcome, the economic consequences of the conflict, not to mention the human losses, have been heavy for both countries.

---

*In the autumn of 1983, France delivered five Super-Etendards to Iraq.

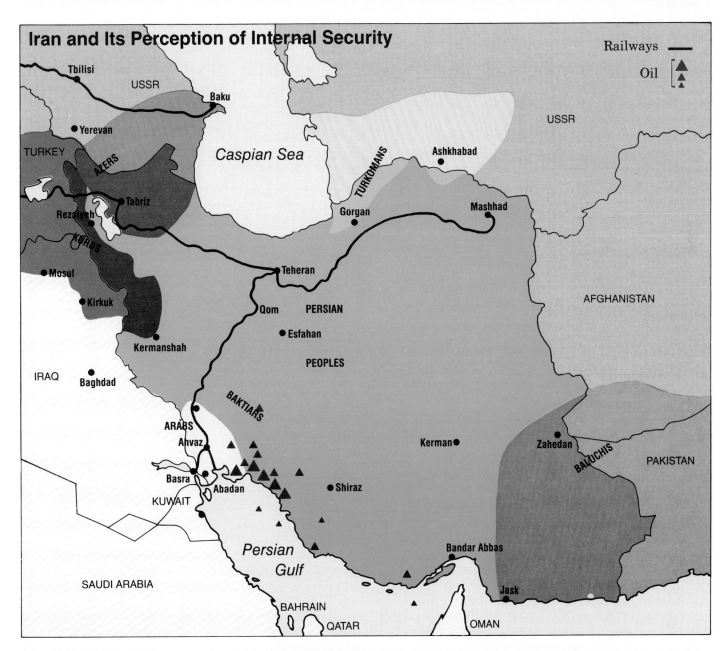

Iran and Its Perception of Internal Security

Railways ▬
Oil ▲

Tbilisi
USSR
Baku
Yerevan
AZERS
TURKEY
Caspian Sea
USSR
Ashkhabad
TURKOMANS
Tabriz
Rezaiyeh
Mashhad
Gorgan
KURDS
Mosul
Teheran
AFGHANISTAN
Qom
PERSIAN
Kirkuk
Esfahan
Kermanshah
PEOPLES
IRAQ
Baghdad
BAKTIARS
Kerman
Zahedan
ARABS
BALUCHIS
Ahvaz
PAKISTAN
Basra
Shiraz
Abadan
KUWAIT
Bandar Abbas
Persian
Gulf
SAUDI ARABIA
Jask
BAHRAIN
QATAR
OMAN

135

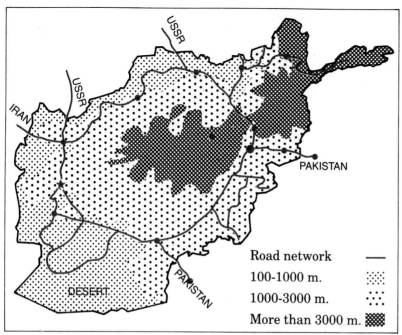

TOPOGRAPHY: AN OBSTACLE TO COMMUNICATIONS

Road network —
100-1000 m. ⠂
1000-3000 m. ⠄
More than 3000 m. ▓

USSR
IRAN
PAKISTAN
PAKISTAN
DESERT

A MOSAIC OF PEOPLES

USSR
CHINA
IRAN
Faizabad
Mazar i Sharif
Maimana
Herat
Kabul
Jalalabad
Peshawar
Shindand
Farah
Kandahar

Baluchis ⠂⠂
Uzbeks ▓
Tadjiks ////
Hazaras ⠂⠂⠂
Pathans ░
Turkomans |||

Afghanistan

The Soviet intervention in Afghanistan is the first Soviet intervention outside the Warsaw Pact countries since the beginning of the Cold War.

The government that emerged from the Marxist-Leninist-inspired coup d'état (1978) rapidly ran up against a massive uprising with many different causes. The Soviet occupation is facilitated by the fact that the two countries share a common border. The USSR's intervention comes on top of Soviet initiatives in Angola (1975) and Ethiopia (1977), which took advantage of strategic vacuums and the weakening of American will after the war in Vietnam and the fall of the Shah (1979).

Afghanistan's traditional buffer-state role is a thing of the past. The USSR has now brought Afghanistan within its camp. Soviet troops will not withdraw from a country deemed to be part of the geostrategic space of the USSR.*

---

*The time when the USSR had to leave Iranian Azerbaijan (1946) is past.

Badly organized but highly motivated, the Afghan rebels, supported by the population, hold the countryside, have external assistance, and all the support required for long-term resistance.

Pakistan is the logistical base of the resistance; it has received 3 million refugees* and is particularly vulnerable, although it is supported by the USA and China as well as by Saudi Arabia.

By occupying Afghanistan, the Soviets have totally transformed Pakistan's security perception, which previously was focused on its rivalry with India.

---

*The official figure of 4 million seems much too high.

Turkey

With a thirty-year lead over Nasserite Egypt, the Turks, after the collapse of an Ottoman Empire undermined by national issues, succeeded, thanks to Mustapha Kemal and a military organization based on a traditional state, in avoiding subjugation. Modern Turkey, the first truly independent nation-state in the Afro-Asian world after Japan, was established. But modernization consisted mostly of the external forms of European institutions and not their economic bases, while the nation remained handicapped by a traditional, "backward" society.

Turkey, a member of NATO, the OECD, and the Council of Europe, is the only state in the Middle East linked to the West by a military pact. Turkey seeks to be European and wants to join the Common Market. In the Middle East its imperial past, its peculiarities, and its political leanings isolate it from its neighbors.

It is the only country in the Middle East outside of Egypt to recognize the state of Israel. Traditionally, the main enemy of the Turks is "the Moscof" (the Russian). The Turkish elites have long been aware of the importance their northern neighbor attaches to the vital geostrategic position of Asia Minor. The fall of the Shah of Iran, which deprived the United States of a major regional ally, increased the value of the alliance with Turkey and the inclination of the West (especially the United States and West Germany) to contribute economic assistance to a country whose stability is necessary to NATO.

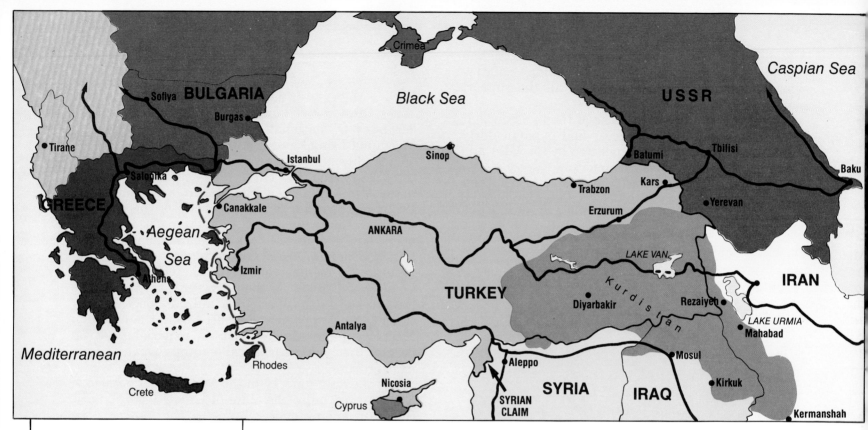

Turkey: Perception of Threats

Legend:

Railways	——
USSR and allies	■
Greece (antagonistic)	■
Arab countries	□
Kurds	■

After the defeat of the Ottoman Empire, a short-lived Armenian republic was set up and recognized within frontiers laid down by President Wilson (in the years 1915–1917, some 1 million Armenians died during deportations ordered by the Young Turks). The Kurds seemed likely to win autonomous status. The Greeks invaded Anatolia.

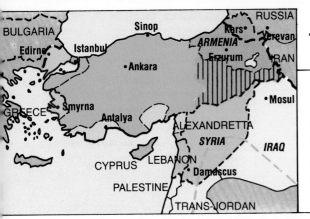

Turkey in 1920

Turkey (Sèvres Treaty 1920)	
Armenia	
Occupied by Greece	
Occupied by France and Italy	
Ceded to France	
Ceded to Britain	
Area planned for Kurdish autonomy	

Birth of Modern Turkey (1919–1922)

Modern Turkey was built on the annihilation of the Armenians and the reconquest of northeastern Anatolia, the expulsion of the Greeks in Asia Minor after their defeat (followed by an exchange of populations), and a process of integration/repression of a large Kurdish minority deprived of all rights as an ethnic group.

The Ottoman Empire in 1914

Cyprus

Turkish intervention in Cyprus (1974), where the Greek (82 percent) and Turkish (18 percent) communities were fighting each other, revived a multifaceted dispute between Greece and Turkey—both members of NATO. This dispute especially concerns Turkish claims to the continental shelf of the Aegean Sea.

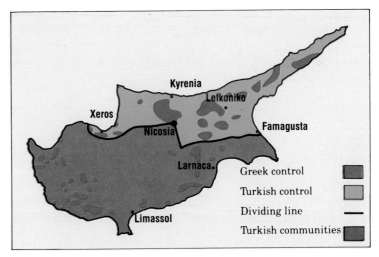

Greek control	
Turkish control	
Dividing line	—
Turkish communities	

Ethiopia, the Muslim World, and the Soviet Presence

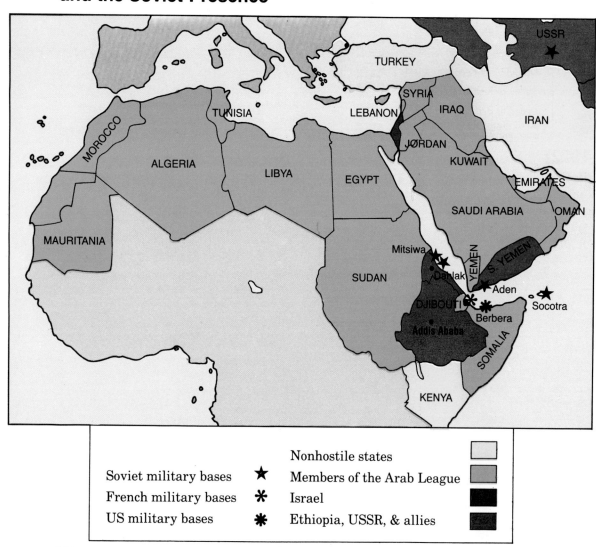

Legend:
- Soviet military bases — ★
- French military bases — ✳
- US military bases — ✴
- Nonhostile states
- Members of the Arab League
- Israel
- Ethiopia, USSR, & allies

Map labels: USSR, TURKEY, SYRIA, LEBANON, IRAQ, IRAN, JORDAN, KUWAIT, EMIRATES, TUNISIA, MOROCCO, ALGERIA, LIBYA, EGYPT, SAUDI ARABIA, OMAN, MAURITANIA, SUDAN, Mitsiwa, Danlak, YEMEN, S. YEMEN, Aden, Socotra, DJIBOUTI, Berbera, Addis Ababa, SOMALIA, KENYA

Ethnic Groups in Ethiopia

While the coastal regions of the Horn were colonized by European powers at the end of the nineteenth century, the Ethiopian state, built on the interior highlands, remained in the hands of Christian Amhara. Taking advantage of inter-European rivalries, Ethiopia succeeded in building up a vast empire which embraced, in the south particularly, non-Christian peoples. The traditional antagonism between Christian Ethiopians and Muslim Somalis was further fueled by the inclusion of the Ogaden, peopled by Somalis, in the empire.

After the overthrow of the Emperor Haile Selassie, who had long been a favored ally of the United States in the region (1974), Ethiopia was radicalized and proclaimed itself Marxist-Leninist.

Confronted by the territorial demands of Somalia (militarily linked by a pact to the USSR), and weakened by civil war, it could win the war in the Ogaden (1977–1978) only with the help of the Soviet about-face toward Somalia and the intervention of Cuban troops.

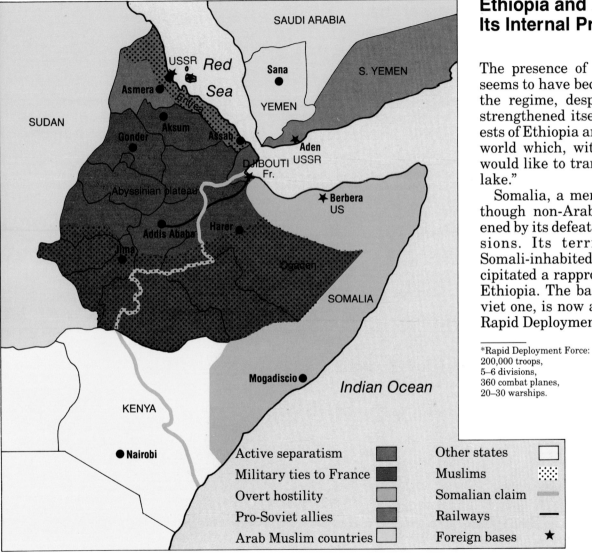

SAUDI ARABIA

USSR **Red**

Sana

S. YEMEN

Asmera

Sea

YEMEN

SUDAN

Aksum

Gonder

Assab

Aden

DJIBOUTI USSR

Fr.

Abyssinian plateau

Berbera
US

Addis Ababa   Harer

Ogaden

SOMALIA

Mogadiscio

Indian Ocean

KENYA

Nairobi

Active separatism		Other states	
Military ties to France		Muslims	
Overt hostility		Somalian claim	
Pro-Soviet allies		Railways	
Arab Muslim countries		Foreign bases	★

Ethiopia and Its Internal Problems

The presence of the Soviets and the Cubans seems to have become permanent, even though the regime, despite the war in Eritrea, has strengthened itself. In the Red Sea, the interests of Ethiopia are opposed to those of the Arab world which, with Saudi Arabia in the lead, would like to transform this sea into an "Arab lake."

Somalia, a member of the Arab League (although non-Arab), has been seriously weakened by its defeats, and suffers Ethiopian incursions. Its territorial claims concerning Somali-inhabited regions in Kenya have precipitated a rapprochement between Kenya and Ethiopia. The base at Berbera, formerly a Soviet one, is now an American base open to the Rapid Deployment Force.*

*Rapid Deployment Force:
200,000 troops,
5–6 divisions,
360 combat planes,
20–30 warships.

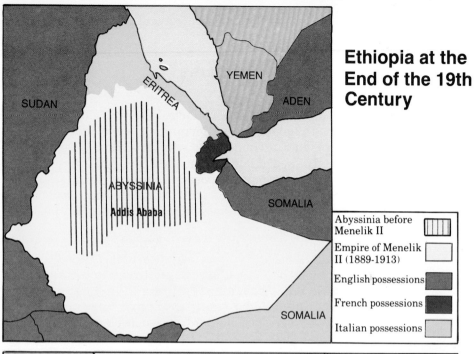

Ethiopia at the End of the 19th Century

SUDAN

ERITREA

YEMEN

ADEN

ABYSSINIA

Addis Ababa

SOMALIA

SOMALIA

Abyssinia before Menelik II	▦
Empire of Menelik II (1889-1913)	☐
English possessions	▨
French possessions	■
Italian possessions	▨

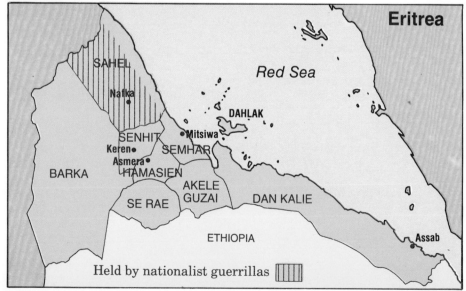

Eritrea

SAHEL

Nafka

Red Sea

DAHLAK

SENHIT

Mitsiwa

Keren

SEMHAR

Asmera

BARKA

HAMASIEN

AKELE GUZAI

SE RAE

DAN KALIE

ETHIOPIA

Assab

Held by nationalist guerrillas ▦

Eritrea

Eritrea was an Italian colony until World War II, then was administered by the British (1941–1952). It did not, like the other Italian colonies, win independence. By a decision of the UN, it was attached to Ethiopia as an autonomous territory (1952). Ten years later, the emperor integrated Eritrea as a province. After 1961, a separatist movement developed, Muslim in origin, but spreading to Christians (each religious group having about half the total population) and supported by the Arab states.

The fall of Haile Selassie did not alter the imperial character of Ethiopia. In spite of Soviet and Cuban aid, the Popular Front for the Liberation of Eritrea (Marxist-Leninist) resisted offensives by Addis Ababa. For Ethiopia, Eritrea represents access to the sea. The USSR has obtained maritime bases in Eritrea. Sudan, formerly openly hostile to Addis Ababa, has to take into account the strengthening of Mengistu's regime and the help given by this regime to the dissidents in southern Sudan (1982).

China and Its Environment

China is a vast country that is strategically isolated. To the north there is the USSR, with which there is a long-standing territorial dispute. In the era of the unequal treaties (those signed to China's disadvantage by various Western countries have long since been nullified), China ceded to Russia in 1853 2,500,000 sq. km. of territory east of the Ussuri as far as the coastal provinces. Absorbed because of their territorial continuity, these regions (which were not inhabited by Chinese) have remained Soviet, but China laid claim to them from the very beginning of the Sino-Soviet conflict.* There was fighting between Chinese and Russians on the Ussuri (1969). After a decade of Sino-American relations, China, disappointed in its hope of obtaining American help in modernizing its military equipment, is seeking an arrangement with the Soviet Union. A normalization of relations would allow it to attend to pressing domestic and economic problems: agriculture, industry, technology, and modernization of the armed forces.

To the south, China is hostile to Vietnamese hegemony over the Indochinese peninsula and would like to see, both in Vientiane and Phnom Penh, governments hostile to the Vietnamese, whose alliance with Moscow is seen as an encirclement. China's only ally in Southeast Asia is a weakened Pakistan. India, beyond the Himalayan barrier, remains hostile.

Heir to a great civilization that has profoundly influenced its neighbors (Vietnam, Korea, Japan) and conquered a vast empire to the west and north of

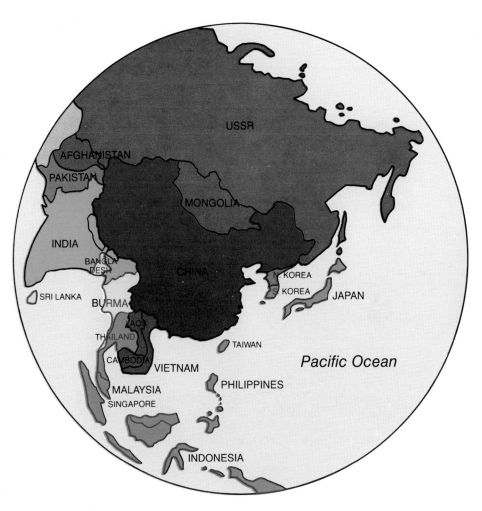

China's original eighteen provinces, the People's Republic has the advantage of a homogeneous population (92% are Han Chinese) with exceptional abilities, whose initial enthusiasm was largely eroded by fifteen years of political struggles and economic stagnation, aggravated by bureaucratic inertia. For the time being, the status of Hong Kong, like that of Macao, is useful to China's trade. Taiwan remains a secondary problem. As for foreign policy, especially regarding the Third World, after numerous years marked by setbacks, it is today conducted on more realistic lines. It is essentially by relying on its own organizational and productive abilities that China, by managing its security through a prudent and adaptable policy, can hope to raise itself, with great effort, to the rank of a great power, a rank that no state is anxious to see it attain.

Cereals*

Wheat	12.5%	(third)†
Corn	11	(second)
Rice	35	(first)
Barley	11	
Millet	36	(first)
Potatoes	5	
Soy	16	(second)
Peanuts	20	(second)
Cotton	19	(second)
Pork	41	(first)

*18% of world production, based on seven principal grains.
†World rank.

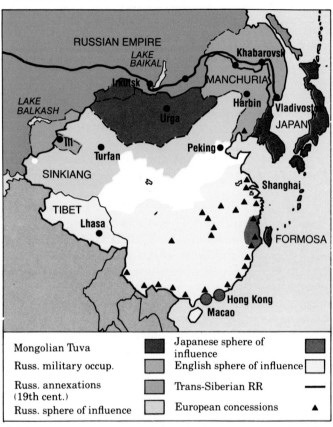

China under Foreign Influence (1904)

Legend:
- Mongolian Tuva
- Russ. military occup.
- Russ. annexations (19th cent.)
- Russ. sphere of influence
- Japanese sphere of influence
- English sphere of influence
- Trans-Siberian RR —
- European concessions ▲

*Underpopulated Mongolia, whose strategic position between China and the USSR is obvious, with its 3,000 km. of frontiers, fears the Chinese, particularly for demographic reasons.

Major Cereals

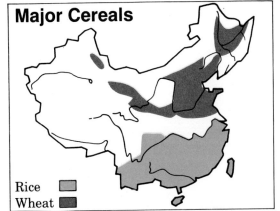

Rice
Wheat

Railways

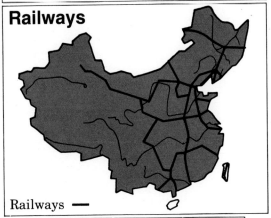

Railways —

Hong Kong

Guangzhou (Canton)

Hong Kong

Victoria

Kowloon

Macao (Port.)

Industrial China

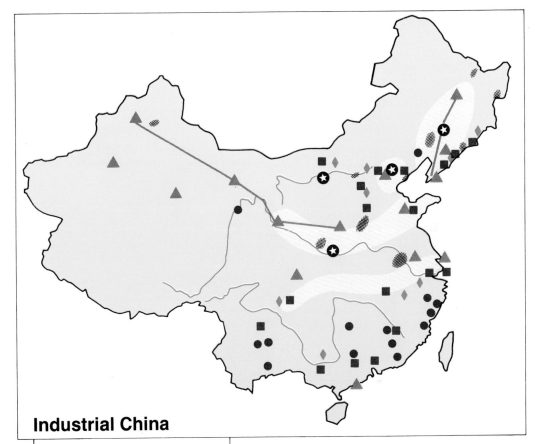

		Underground Resources	World Production	Reserves
Industrial areas	▢	Coal and lignite	19 %	28 %
Industrial centers	■	Petroleum	3	3
Nuclear power plants	✪	Iron	8.3	4
Iron	◆	Boron	— *	38
Other ores	●	Manganese	6	— *
Oil and natural gas	▲	Tungsten	26	47
Coal	▨	Vanadium	14	13
		Antimony	18	39

*No data.

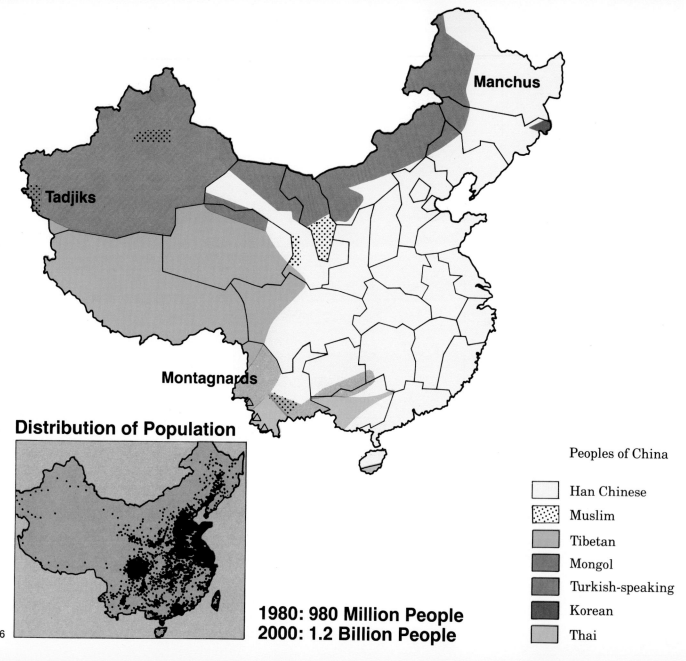

Manchus

Tadjiks

Montagnards

Distribution of Population

1980: 980 Million People
2000: 1.2 Billion People

Peoples of China

- Han Chinese
- Muslim
- Tibetan
- Mongol
- Turkish-speaking
- Korean
- Thai

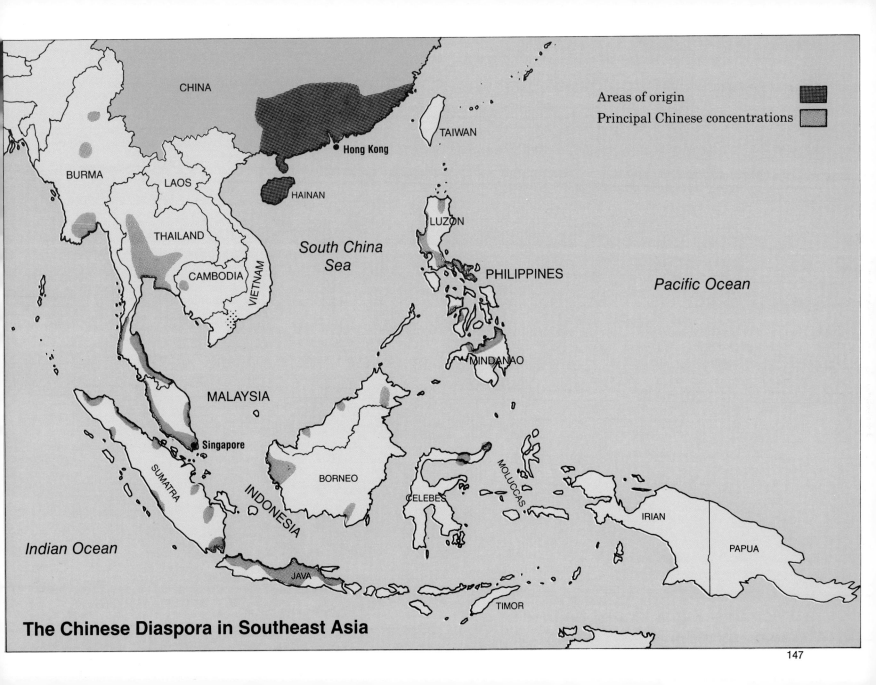

The Chinese Diaspora in Southeast Asia

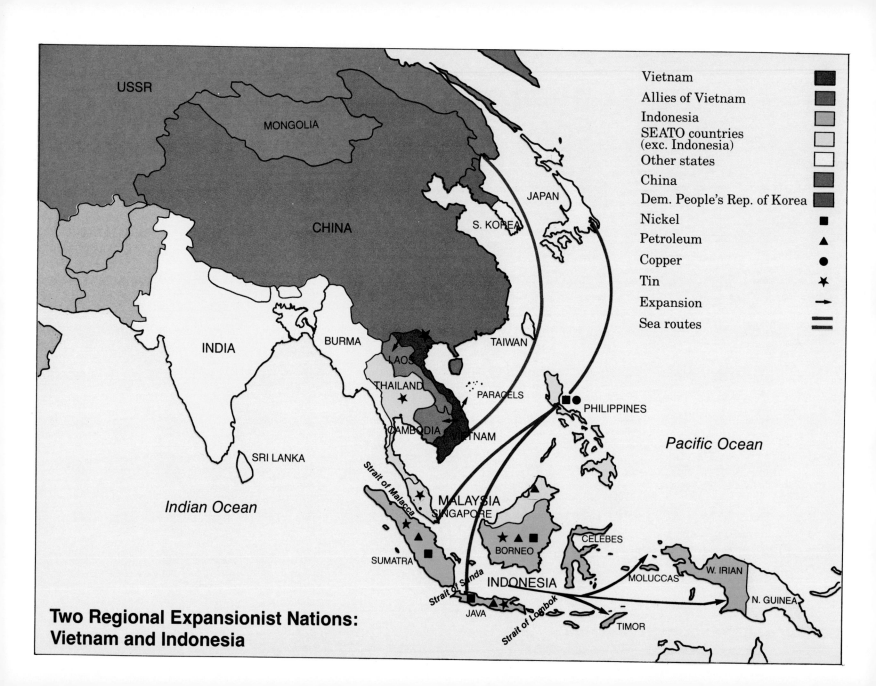

**Two Regional Expansionist Nations:
Vietnam and Indonesia**

USSR

MONGOLIA

CHINA

JAPAN

S. KOREA

INDIA

BURMA

TAIWAN

LAOS

THAILAND

PARACELS

CAMBODIA   VIETNAM

PHILIPPINES

Pacific Ocean

Indian Ocean

SRI LANKA

Strait of Malacca

MALAYSIA
SINGAPORE

SUMATRA

BORNEO

CELEBES

W. IRIAN

Strait of Sunda

INDONESIA

MOLUCCAS

N. GUINEA

JAVA

Strait of Lombok

TIMOR

Vietnam

Allies of Vietnam

Indonesia

SEATO countries
(exc. Indonesia)

Other states

China

Dem. People's Rep. of Korea

Nickel ■

Petroleum ▲

Copper ●

Tin ★

Expansion →

Sea routes ═

Southeast Asia

Monsoon Asia is predominantly made up of peninsulas and islands surrounded by sea lanes. Like the whole of eastern Asia, it is an area of high agricultural production (rice) and dense population (except Laos). Without ethnic or religious unity (Hinduism, Buddhism, Islam, Confucianism, and Catholicism), Southeast Asia is characterized by societies that are highly structured, with a national consciousness that is often very old and with considerable cultural depth.

The victory of the North Vietnamese in 1975 and the intervention in Cambodia (1978) signified the hegemony of Vietnam over the Indochinese peninsula, provoking the militant hostility of China. This resulted in an armed Sino-Vietnamese confrontation (1979) and in the support given by Thailand and Malaysia (both members of ASEAN*) to the various Khmer forces fighting the Hanoi-backed regime in Cambodia.

The withdrawal of Vietnamese forces, whose military superiority is quite apparent, is out of the question so long as there is no guarantee that Cambodia will be led by a pro-Vietnamese government. Chinese power is perceived as a threat by the Vietnamese, who feel that their hegemony over the Indochinese peninsula gives them a status as a regional power that cannot be ignored.

China in turn feels threatened by the alliance between the USSR and Vietnam. Despite the existence of chronic guerrilla wars (Thailand, Malaysia, Burma, Cambodia, Laos, the Philippines), no existing regime appears threatened by them. Economically, the most dynamic states are Singapore, Thailand, and Malaysia. But it is interesting to note that the two regional powers that have made territorial advances are Vietnam and Indonesia (Moluccas, 1950–1952; New Guinea, 1961–1962; East Timor, 1976–1977); Indonesia also has designs on North Borneo.

*Association of Southeast Asian Nations, founded in 1967. This economic and political organization comprises Indonesia, the Philippines, Singapore, and Thailand.

Population (in millions)

	1950	1980	2000
Indonesia	77	148	216
Malaysia	6.1	14	21
Philippines	20	49	77
Singapore	1	2.4	3
Thailand	20	47	68
Vietnam	30	55	88
Cambodia	4	6.9	10
Laos	2	3.4	5
Burma	19	35	54
Hong Kong	2	5.1	6
Taiwan	8	18	25

East Asia: One of the Rare Third World Regions with Sustained and Rapid Growth

	1980	(annual % rate) 1985 (prediction)
South Korea	3.5	6.0
Hong Kong	9.0	9.6
Indonesia	9.3	9.6
Malaysia	8.0	9.5
Philippines	5.2	8.2
Taiwan	6.4	8.0
Thailand	6.3	8.0

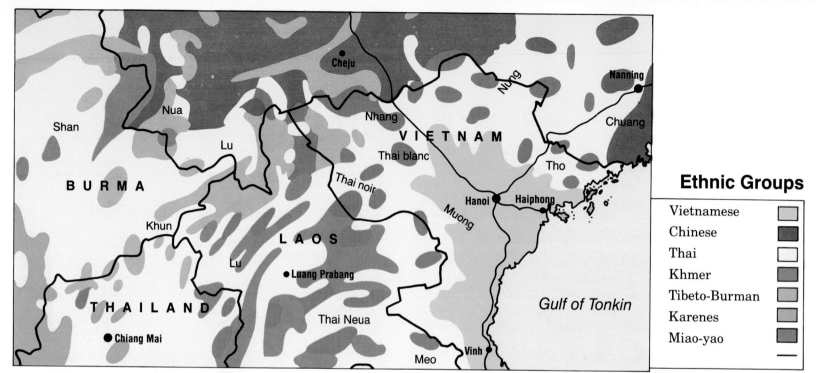

Ethnic Groups

Vietnamese	
Chinese	
Thai	
Khmer	
Tibeto-Burman	
Karenes	
Miao-yao	

Labels on map: Cheju, Nanning, Nua, Shan, Nhang, Nung, Chuang, Lu, VIETNAM, Tho, Thai blanc, BURMA, Thai noir, Khun, Muong, Hanoi, Haiphong, LAOS, Lu, Luang Prabang, Gulf of Tonkin, THAILAND, Thai Neua, Chiang Mai, Meo, Vinh

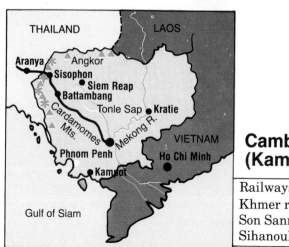

Cambodia (Kampuchea)

Labels on map: THAILAND, LAOS, Aranya, Angkor, Sisophon, Siem Reap, Battambang, Cardamomes Mts., Tonle Sap, Kratie, Mekong R., VIETNAM, Phnom Penh, Ho Chi Minh, Kampot, Gulf of Siam

Railways	—
Khmer rouge	▲
Son Sann forces	✳
Sihanouk forces	★

An Area of Geopolitical Fluidity

On the borders of China and several Southeast Asian countries, the profusion of minorities, their overlapping, and the fact that they are geographically exclusive makes the geopolitical situation particularly fluid. Political guerrillas that are more or less manipulated by one power or another (China, USA, etc.) and bands of irregulars engaging in the opium trade continue to find rear bases or sanctuaries in these areas.

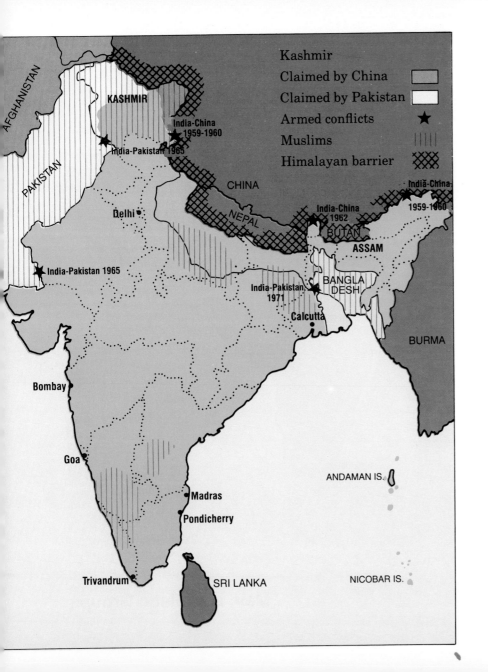

Kashmir
Claimed by China
Claimed by Pakistan
Armed conflicts ★
Muslims
Himalayan barrier

India: Its Security Perception

India is, with China, the only great regional power in Asia. India, a quasi-industrial power that belongs to the nuclear club (1974), has more to fear from its internal disparities and distortions, fueled by an almost uncontrollable population growth, than from its neighbors.

India is not a nation, but a civilization. Since independence, the language of administration has been English. In addition to its linguistic and ethnic divisions (especially north–south), there is the problem of its very large Muslim minority (almost 20%). Social cleavages, partly rigidified by the caste system, exclude 105 million Harijans (formerly Untouchables), despite legislation.

Twice defeated by China (1959, 1962), India has grown stronger since, and China seems unlikely in the foreseeable future to embark on incursions that, in any case, would be logistically difficult to make on a large scale.

Pakistan, which claims Muslim-inhabited Kashmir, is today, after many conflicts, greatly weakened by its defeat in Bangladesh by Indian troops and by the closeness, on the Afghan frontier, of Soviet troops. India has always had excellent relations with the USSR (especially in the matter of military supplies) while remaining a model of nonalignment. However, the excessive proximity of Soviet troops might worry a political elite that is skillful and conscious of its interests.

The overseas Indian population is numerous throughout eastern Africa, from Kenya to South Africa, as well as in the eastern part of the Indian Ocean (Malaysia, Singapore, etc.). India is one of the few Third World countries to have the means to carry out its south–south industrial projects. Given its ambitions in the Indian Ocean area, it would seem that strengthening its maritime capabilities, which are already quite considerable, is necessary.

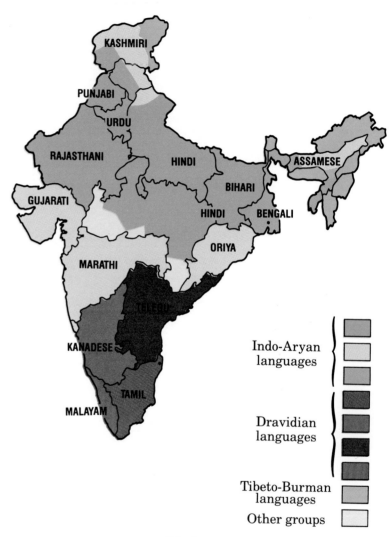

Indo-Aryan languages

Dravidian languages

Tibeto-Burman languages

Other groups

Main Linguistic Groups

Economic India

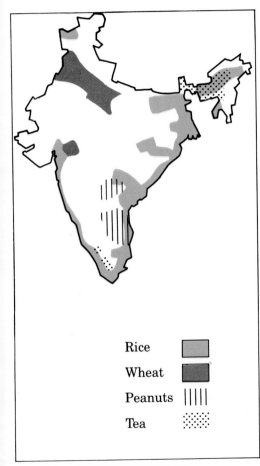

MAIN AGRICULTURAL RESOURCES

Rice

Wheat

Peanuts |||||

Tea

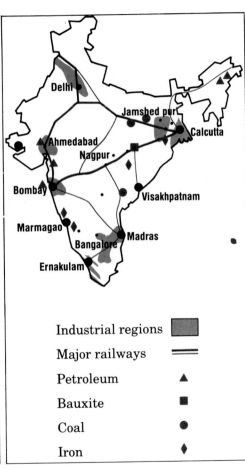

INDUSTRY

Industrial regions

Major railways

Petroleum ▲

Bauxite ■

Coal ●

Iron ◆

Cereals—9% of world production, on the basis of principal cereals

Rice	20%
Wheat	7%
Millet	30%
Cassava	6%
Sugar	5%
Potatoes	4%
Peanuts	34%
Cotton	10%

Mineral and Energy Resources

	World Production (%)	
Coal	3%	
Iron ore	5%	(5.5% of reserves)
Manganese	6%	
Ilmenite (titanium)	20%	(reserves)
Thorium	30%	(reserves)

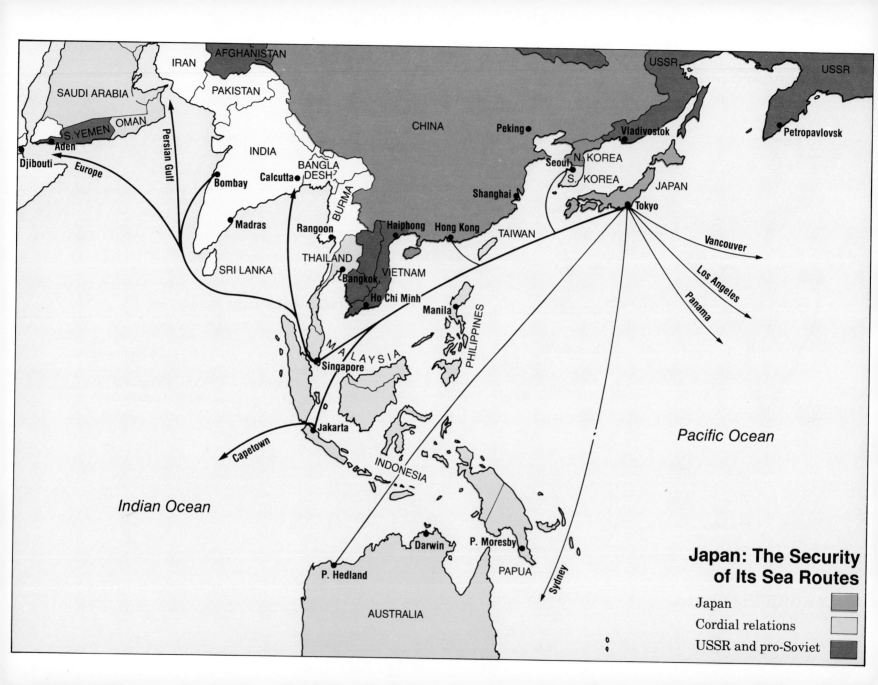

Japan: The Security of Its Sea Routes

Japan

Cordial relations

USSR and pro-Soviet

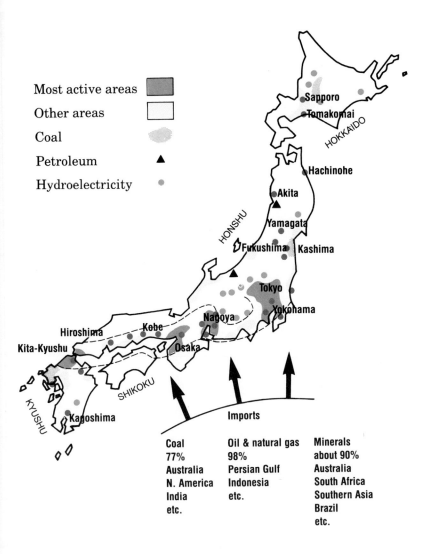

Most active areas

Other areas

Coal

Petroleum ▲

Hydroelectricity ●

Sapporo
Tomakomai
HOKKAIDO

Hachinohe

Akita
Yamagata
Fukushima   Kashima
HONSHU

Tokyo
Yokohama
Nagoya

Kobe
Hiroshima
Kita-Kyushu   Osaka
SHIKOKU

KYUSHU   Kagoshima

Imports

Coal 77%	Oil & natural gas 98%	Minerals about 90%
Australia	Persian Gulf	Australia
N. America	Indonesia	South Africa
India	etc.	Southern Asia
etc.		Brazil
		etc.

Japan: An Industrial Power

Japan is the only Afro-Asian country that succeeded in responding to the Western challenge and in carrying out an industrial revolution launched in 1868. Since then, it has spared no effort to maintain its independence. At the end of a war in which it had imposed its harsh law on East Asia, Japan suffered nuclear tragedy (Hiroshima, Nagasaki).

Provided with democratic institutions during the American occupation, demilitarized* Japan reoriented its energy, organizational abilities, cohesion (founded, among other things, on a population whose homogeneity has been carefully preserved), and social discipline toward economic and commercial goals. Lacking resources, and aided by a dynamism backed by modern business organization and the consensus of wage earners, Japan has succeeded, in three decades, in outclassing almost all the industrial countries, becoming an irresistible competitor. Militarily, Japan is highly vulnerable, but its economic vulnerability is even greater, given the country's heavy dependence on the world market: A break in supplies of raw materials from the Third World or the erection of barriers in the industrial countries where it sells its products, would be serious. Japan's competitiveness tends to provoke protectionist measures in some countries. Japan's security, at the present time, rests on the continuation, as far as it can be done, of free trade.

---

*Japan is linked militarily to the United States.

Having succeeded in controlling its birth rate in the 1950s, enjoying a productive agriculture, engaging in large-scale fishing, and endowed with a hard-working and docile population, Japan has the necessary determination to face all challenges. Its lines of communication—which are vital, especially from the Gulf—are guaranteed by a large merchant fleet. In the framework of new rules of the game to which it has adapted, Japan seems to have realized many of the objectives it had during the years of empire.

Urbanized Zones in the South

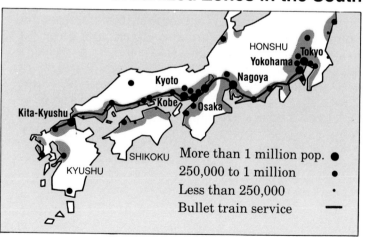

More than 1 million pop. ●
250,000 to 1 million ·
Less than 250,000 ·
Bullet train service ▬

Population Density

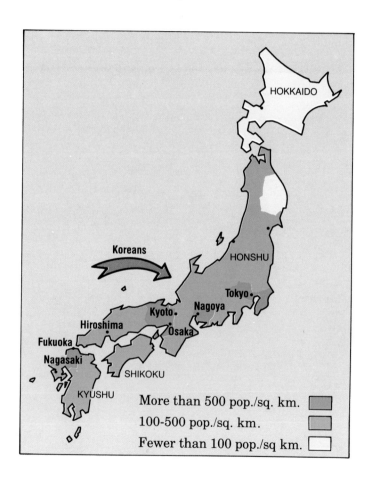

More than 500 pop./sq. km.
100-500 pop./sq. km.
Fewer than 100 pop./sq. km.

Industrial Power: % of World Production

Production

Steel	14.0
Copper	11.5
Aluminum	7.0
Shipbuilding	54.0
Autos	24.5
Commercial vehicles	42.0
Resins and plastics (petrochemicals)	14.0
Synthetic rubber	11.0
Synthetic fabrics	13.0

Resources

Foodstuffs

Rice	approx. 13 million tons
Fish	15% (of world total)

Fuels—Ores

Coal	12% of total consumed
Hydroelectricity	13% of total electricity consumed

Imports (Dependencies)

Wheat, cereals, agric. products	25% of needs
Copper	100% of needs
Coal	88% of needs
Oil	99% of needs
Uranium	100% of needs
Iron	92% of needs
Other ores	+90% (except lead, zinc, silver)

Japanese Foreign Trade 1981 (%)

Exports		Imports
25	United States	18
16	Western Europe	8
13	EEC	6
15	OPEC	38
20	Far East	17
24	Others	19
$152	Total value (billions of dollars)	$143

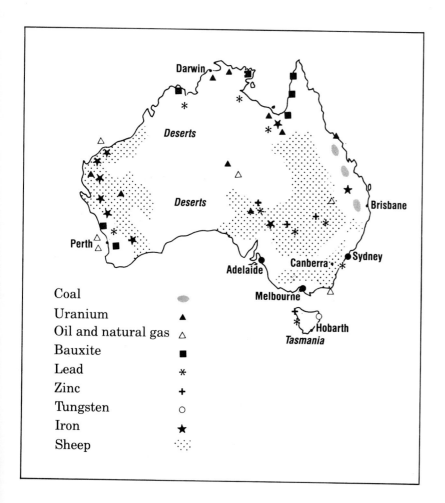

Coal

Uranium ▲

Oil and natural gas △

Bauxite ■

Lead *

Zinc +

Tungsten ○

Iron ★

Sheep ∴

Australia

Australia cooperates militarily with the United States in ANZUS (Australia, New Zealand, United States) as well as with Singapore and Malaysia in the framework of the five-nation agreement (Great Britain, Australia, New Zealand, Singapore, Malaysia). It participates in the sea, air, and naval forces detached by the American Seventh Fleet in the Indian Ocean.

Australia occupies an exceptional geostrategic position between the Indian Ocean and the Pacific. As the largest state in the Pacific, it seeks to play a role as a regional power. Its mineral resources are considerable: It has offshore oil; in addition, it is a leading world exporter of grain and cereals.

A close trading partner of Japan, to which it supplies many raw materials, Australia seeks to work with New Zealand, which plays a more modest role.

Sheep	12% (world flocks) 2nd rank	
Uranium	3.5% world prod.	7% reserves
Iron	10.5%	11%
Manganese	7.5%	6%
Tungsten	6.0%	4%
Cobalt	6.0%	11%
Bauxite	28.5%	20%
Lead	11.5%	14%
Zinc	8.5%	10%

Australia: Perception of Its Security

Australia's traditional sensitivity to possible threats from Asia is fueled by awareness of its isolation in the southern hemisphere and its low population.

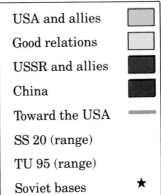

USA

CANADA

San Francisco
6450 naut. miles

USSR

MONGOLIA

HAWAII

INDIA

CHINA

KOREA

JAPAN

TAIWAN

Pacific Ocean

THAILAND

PHILIPPINES

VIETNAM

SS-20

MALAYSIA

ASEAN   INDONESIA

Darwin

TU-95

Indian Ocean

AUSTRALIA   Brisbane

Perth

Sydney

Melbourne

Auckland

NEW ZEALAND

USA and allies	
Good relations	
USSR and allies	
China	
Toward the USA	
SS 20 (range)	
TU 95 (range)	
Soviet bases	★

Latin America

Latin America is economically dominated by the United States, to which it is linked by the Treaty of Rio (1947), an inter-American mutual defense treaty, and by the Organization of American States (1948), from which Cuba was excluded in 1962 at the time of the missile crisis.

The United States historically has defended its geostrategic area: Guatemala (1954), Bay of Pigs (1961), and, following Cuba's move to the left, through the Alliance for Progress, a combination of economic assistance and training of Latin American counterguerrilla forces. (All the Latin American guerrilla movements of the 1960s have been defeated: Brazil [coup d'état, 1964]; intervention in the Dominican Republic [1965]; destabilization and fall of the Allende government in Chile [1973].)

Basically, the Latin American world is more stable than it appears: In three decades, there have been only two radical political changes—Cuba (1959) and Nicaragua (1979).

Latin America has seen exceptionally rapid population growth: 132 million in 1945; 303 million in 1975; 600 million projected for 2000. The dominance of the Spanish language (except in Brazil) throughout Latin America should not lead one to overlook the existence of large In-

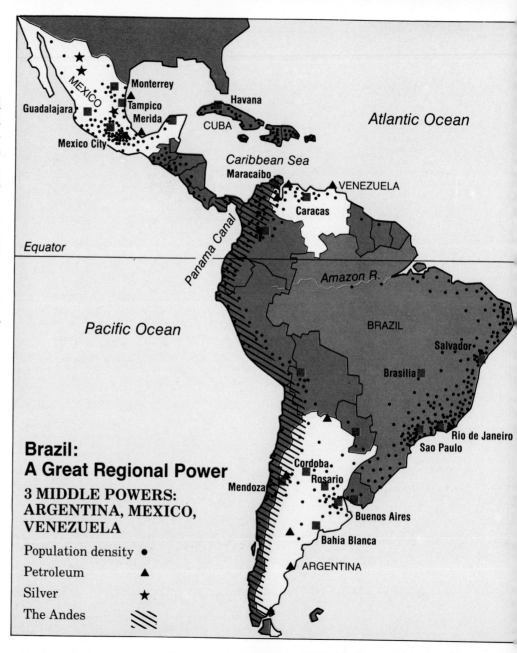

**Brazil:
A Great Regional Power**

**3 MIDDLE POWERS:
ARGENTINA, MEXICO,
VENEZUELA**

Population density   ●

Petroleum   ▲

Silver   ★

The Andes   ⫰⫰⫰

dian populations (Bolivia, Peru, Ecuador, Guatemala, Honduras, Mexico, etc.), almost all of whom are marginal people in their societies. Outside Brazil, the black or colored populations are concentrated in the Caribbean basin.

There has been no integration or resolution of the crisis-laden social distortions, despite strong nationalist feelings and social problems to which the Catholic Church has given expression. Catholicism is a force to be reckoned with; at the end of the century, the majority of Catholics in the world will be Latin Americans.

The continent has substantial resources, although only one country is in a position to export a surplus in agriculture or livestock (Argentina), and only one has really large mineral resources (Brazil). Oil, in modest quantities, is exploited in Mexico, Venezuela, and Ecuador. Peru has five major minerals, but Chile (copper), Bolivia (tin), and Jamaica (bauxite) have only one each. Although backward areas, notably among the Andean countries and in Central America, would seem to have limited prospects, Latin America as a whole is in a much better situation than most of Asia and Africa.

The present crisis and the wars in Central America are perceived by the United States as a test of political will, and it seems that everything will be done to ensure that the Pax Americana prevails in the end.

Three decades ago, Argentina seemed, given its cultural level and its agricultural and cattle wealth, to be destined to become a major power. But neither in terms of development, population, or institutions has Argentina lived up to its promise. The failure of the Falklands gamble, underestimating the capacity of Mrs. Thatcher's government to respond, has accentuated the crisis in Argentine society. The relations of convenience between right-wing soldiers and the USSR since 1980, based on common interests, is viewed with annoyance in Washington, which was obliged to support Great Britain during the Falklands crisis.

Venezuela, with a small population, can lay claim to the role of middling regional power only because of its oil. Its geostrategic position is linked to the Caribbean as a whole, and its interest lies in maintaining stability there. The 200-nautical-mile limit would allow Venezuela to exercise rights over a significant part of the Caribbean. The drawing of the limits of the continental shelf raises problems with its rival Colombia, which exports a large part of its work force to Venezuela.

Mexico, once considered a rising regional power, demonstrated its fragility in 1982 with the collapse of its currency. Contrary to what it has been claiming for several decades, the Institutional Revolutionary Party (PRI), which has been in power for half a century, has not succeeded in modernizing the country's institutions. Corruption, which is typical of the system, has increased with the oil boom. Today, Mexico is faced with a crisis that may well have serious social consequences.

Brazil: South–South Perception

The Organization of American States (OAS), formed in 1948, reaffirmed the cooperation among the American nations (the First Congress of American States, 1889), as did the reciprocal-aid Treaty of Rio, signed in 1947. Members include: All the South American states except British and French Guyana; the Antilles, except for Cuba, which has not participated, including Jamaica, Haiti, the Dominican Republic, Antigua and Barbuda, Dominique, St. Lucia, St. Vincent and the Grenadines, Trinidad, and Tobago. Different economic alliances regroup some of these countries, such as the Andean group, countries of the Caribbean basin, and the Association for Latin American Integration (ALADI).

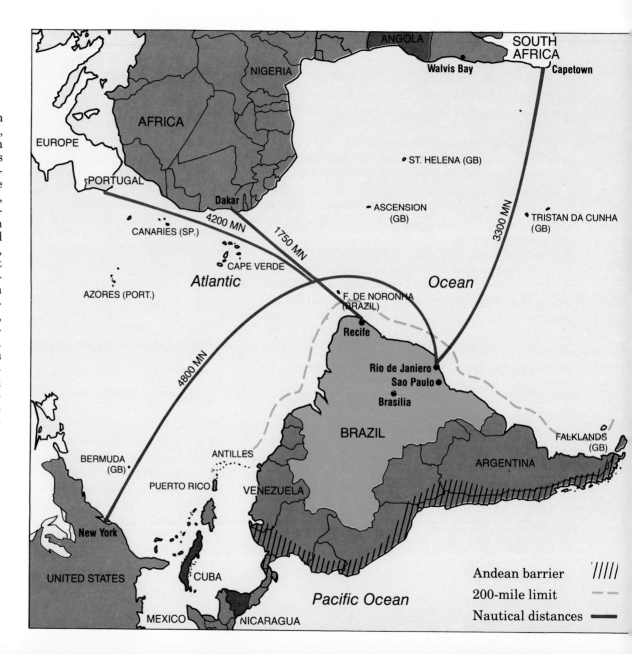

Brazil: Occupation of the Land

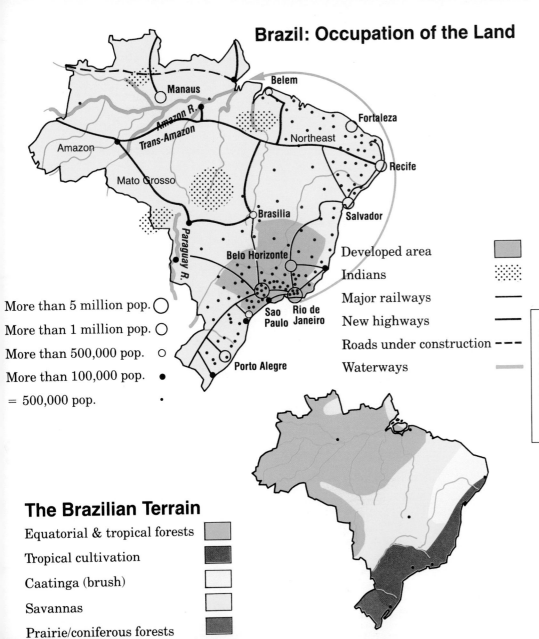

Corn	6%
Cassava	30%
Soya	17%
Sugar	10%
Beef	7%
Pork	5%

More than 5 million pop. ◯

More than 1 million pop. ◯

More than 500,000 pop. ○

More than 100,000 pop. ●

= 500,000 pop. ·

Developed area

Indians

Major railways

New highways

Roads under construction

Waterways

Hydroelectricity	6% (world prod.)
Iron	11.5% (world prod.) 20% res.
Manganese	8% (world prod.)
Niobium	87% (world prod.) 23% res.
Bauxite	5% (world prod.) 10.5% res.
Rutile	74% res.
Thorium	6% res.
Tantalum	25.5% (world prod.)

The Brazilian Terrain

Equatorial & tropical forests

Tropical cultivation

Caatinga (brush)

Savannas

Prairie/coniferous forests

163

Brazil and Its Resources

Brazil, whose perception of the world is south–south oriented, is like a promontory at the eastern tip of the South American continent, less than 3,000 km from Dakar. It is the only Portuguese-speaking state in Latin America; it is also the most populous and is the only regional power. Moreover, it has considerable ambitions, but the economic base that would enable them to be realized is still lacking.

The development of Brazil depends on the conquest and control of its own territory and its ability, which is currently uncertain, to maintain dynamic growth.

But whatever the difficulties, Brazil is lucky to be able to count on a vast quantity of unexploited resources and is better placed than others to pursue a go-ahead policy.

A multiracial society based on "cordial domination" and mixing may, despite appearances, be the scene of racial problems that are also social problems.

Once tempted to base its African policy on South Africa, Brazil has reoriented its relations toward the Portuguese-speaking states and Nigeria. The creation of a navy suited to its ambitions should be a priority.

Intensive agriculture

Extensive agriculture

Coal

Hydroelectricity

Petroleum

Nickel

Iron

Manganese

Copper

Nuclear power plants

Hydroelectric systems

Refineries

Storage facilities

Electric networks

Pipelines

THE INDUSTRIALIZED CENTER

Brazil and Its Geopolitical Aspirations

*Portugal and its colonies occupy an enviable situation
in the world beyond Latin America that can never be
adequately stressed. In both the North Atlantic, where
the Azores, Madeira, and Cape Verde constitute un-
paralleled defensive outposts, and in the south of Africa, where Angola and Mozambique almost mark out
a Lusitanian equator right opposite the main power
center that we in Brazil represent, and that is without
mentioning Guinea, a second Dakar. . . .**

*By its prominent situation in the nearest semicircle,
which is vital for South America and Brazil's security,
this area creates a Portuguese responsibility that we
must be ready to take on at any moment.*

*The Latin world, in its turn, though its ties are
looser, must consciously accept a sphere of solidarities
that has come to include a large part of the European
peninsula and almost the whole of West Africa. And
this is because we are a Latin country, by our origin
and by our culture, with an outstanding and hard-
working population.*

(Golbery do Couto e Silva, *Conjuntura politica nacional o poder
executivo e geopolitico do Brasil*, Livraria José Olympio editore,
Rio, 1981, p. 195; from the French translation by Alain
Mangin.)

---

*Written in 1959.

Central America: Zone of Conflicts

Central America, divided into a group of micro-states, is a geographical area traditionally controlled by the United States, which has intervened on numerous occasions from 1903 (Panama) to 1965 (Dominican Republic). Central America is currently an arena of East–West confrontations arising out of local conflicts.

The situations in Nicaragua, El Salvador, Guatemala, and Honduras are interdependent. The United States is aiding El Salvador and Guatemala in their fights against insurrections and is using Honduras as a base, particularly to shore up the anti-Sandinista movements in Nicaragua.

Nicaragua, allied to the Cubans and the Soviets, is assisting the guerrillas in El Salvador. The armed struggles in Guatemala (in which Indians are taking part) and in El Salvador, Marxist in inspiration, are well organized. But the United States cannot accept their successes and the spread of revolutionary movements in the region.

The zones held by the Salvadoran guerrilla movement, Morazan and Chalatenango, are poor and quite densely populated, despite the exodus caused by the war. The country is increasingly cut in two, but it does not seem likely that either side can win a military victory.

Guatemala, richer and more populous, is having to face an insurrection in which a large Indian population is taking part. This fact is perhaps the major new feature of the decade.

Panama: The Canal Zone

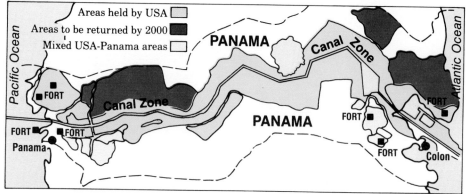

Areas held by USA
Areas to be returned by 2000
Mixed USA-Panama areas

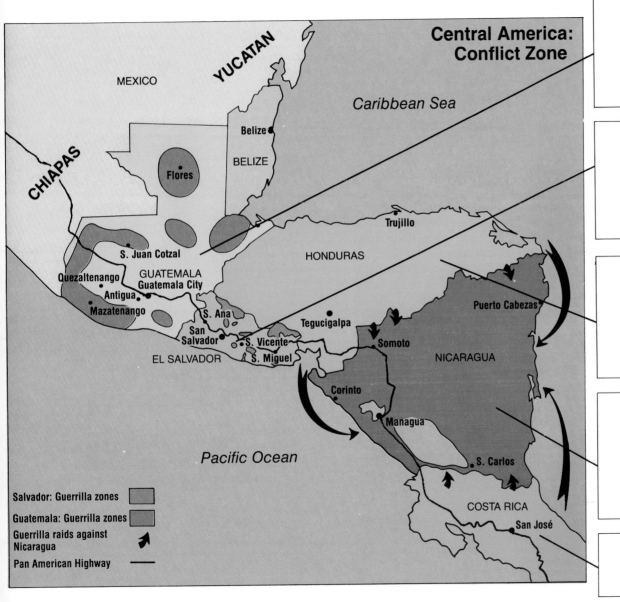

Central America: Conflict Zone

YUCATAN

MEXICO

CHIAPAS

Caribbean Sea

Belize

BELIZE

Flores

S. Juan Cotzal

Trujillo

HONDURAS

Quezaltenango

GUATEMALA
Guatemala City

Antigua

Mazatenango

Puerto Cabezas

S. Ana

Tegucigalpa

San Salvador

S. Vicente

Somoto

EL SALVADOR

S. Miguel

NICARAGUA

Corinto

Managua

Pacific Ocean

S. Carlos

COSTA RICA

San José

Salvador: Guerrilla zones

Guatemala: Guerrilla zones

Guerrilla raids against Nicaragua

Pan American Highway

Guatemala
109,000 sq. km.; pop. 7 million
Indians: 55 to 60% of population
Urban population: 30%
Security forces: 45,000 men
Army: 23,000 men
Guerrillas: 4,000 to 6,000 men

El Salvador
21,000 sq. km.; pop. 4.5 million
Indians: 55 to 60% of population
Urban population: 40%
Security forces: 25,000 men
Army: 16,000 men
Guerrillas: 5,000 to 7,000 men
American advisors: 55

Honduras
112,000 sq. km.; pop. 3.7 million
Indians: 30 to 35% of population
Urban population: 35%
Security forces: 25,000 men
Army: 15,000 men
American advisors: 120
Guerrillas: several hundred

Nicaragua
130,000 sq. km.; pop. 2.6 million
Indians: 5 to 10% of population
Urban population: 57%
Security forces: 16,000 men
Army: 25,000 men
Soviet or Cuban advisors: 2,000
Anti-Sandinista commandos: 8,000 to 9,000 men

Costa Rica
51,000 sq. km.
Population: 2.2 million
Urban population: 46%

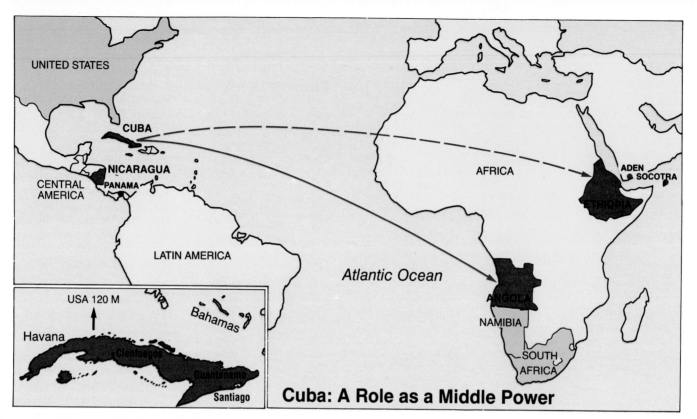

Cuba: A Role as a Middle Power

Despite its small size, Cuba acts as a middle power, but this role is heavily dependent on Soviet logistical support.

Even though the Cuban presence exists in a reduced form in a number of African states, it directly influences the political situation only in Angola and Ethiopia as a result of the massive intervention of Cuban troops in 1975 and 1977.

For a nation of some 10 million inhabitants, Cuba has, with about 25,000 soldiers, proportionately a very high presence in Africa (for the USA, it would equal nearly 600,000 men).

After its setbacks in Latin America in the 1960s, Cuba has regained an active role there with the Sandinista government in Nicaragua and now has influence over the conflicts in Central America (El Salvador, Guatemala).

Significant Cuban Presence:

Angola: 12,000–18,000
Ethiopia: 7,000–9,000
Other African countries: More than 3,000, including 1,000 in Mozambique and 800 in the Congo.

NATURAL CONSTRAINTS

Natural constraints determine the density of human settlement and the eventual exploitation or development of mineral or agricultural resources. They may be obstacles or advantages in times of war and guerrilla activity. Nature, as a constraint and as a resource, is an element of strategy.

Basic Data

World Land Use (%)

	Cultivated Land	Pasture	Forest	Uncultivated Land
USSR	10%	17%	41%	32%
N. America	14%	13%	34%	39%
S. America	7%	26%	51%	16%
Africa	8%	30%	23%	39%
Middle East	7%	16%	12%	65%
Asia (exc. USSR & Middle East)	33%	4%	40%	23%
Europe (exc. USSR)	26%	19%	33%	22%
Oceania	6%	60%	18%	16%

Forests (in millions of hectares)

Surfaces Covered	1978	2000
USSR	785	775
North America	470	464
South America	550	329
Africa	188	150
Asia (except USSR)	361	181
Europe (except USSR)	140	150
Japan		
New Zealand		
Australia	69	68
Total	2,563	2,117

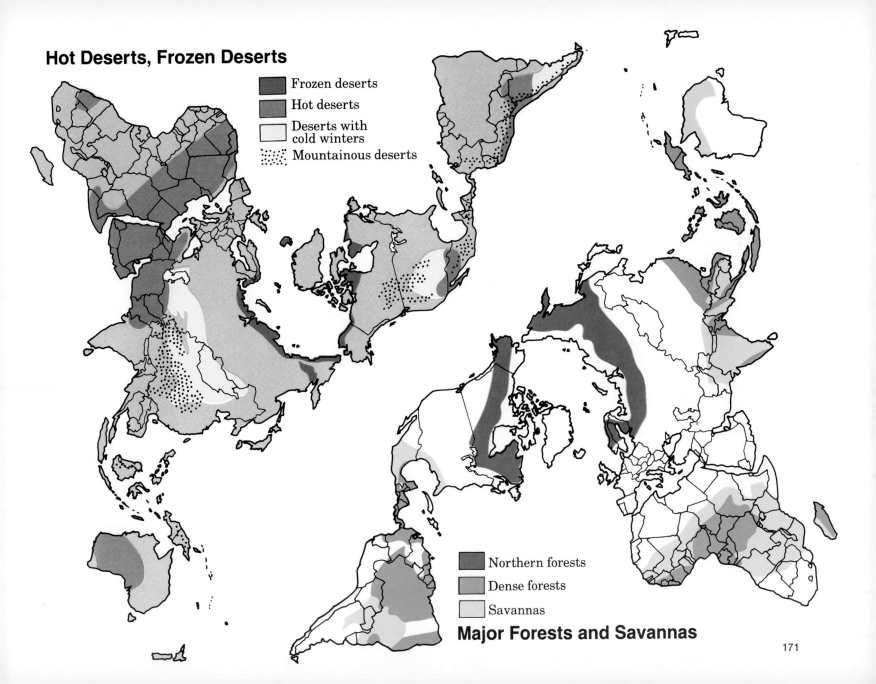

Hot Deserts, Frozen Deserts

- Frozen deserts
- Hot deserts
- Deserts with cold winters
- Mountainous deserts

- Northern forests
- Dense forests
- Savannas

Major Forests and Savannas

171

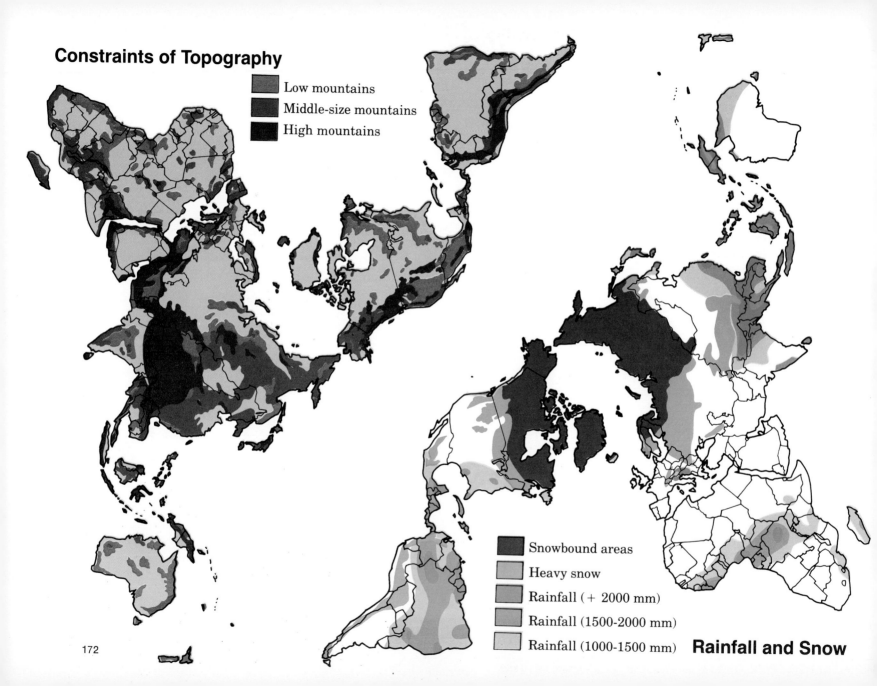

Constraints of Topography

- Low mountains
- Middle-size mountains
- High mountains

Rainfall and Snow

- Snowbound areas
- Heavy snow
- Rainfall (+ 2000 mm)
- Rainfall (1500-2000 mm)
- Rainfall (1000-1500 mm)

172

Availability of Drinking Water

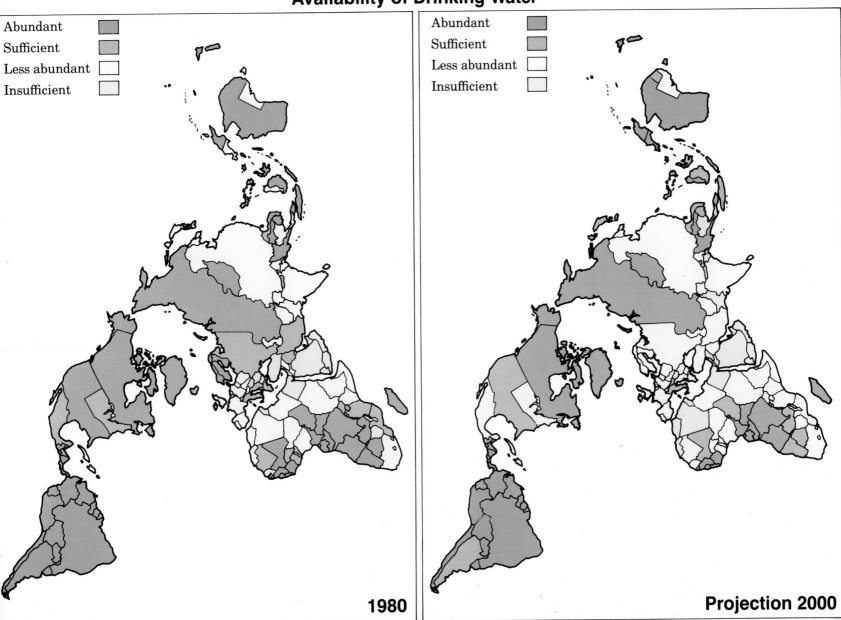

Abundant
Sufficient
Less abundant
Insufficient

1980

Abundant
Sufficient
Less abundant
Insufficient

Projection 2000

Source: *The Global 2000 Report to the President,* Washington, DC, 1980.

ECONOMIC DATA

Rich and Poor in Minerals

The main areas of mineral production and most known reserves are situated in the great industrial states of the northern hemisphere, in South Africa, and in Australia. Western Europe and Japan have few of them.

Gold production
South Africa   51%
USSR   31.5%
Canada   3%

Mineral products
equal more than 5%
of world total.

More than 6 minerals
DEVELOPED STATES
Poorly supplied

More than 2 minerals
THIRD WORLD
STATES
Poorly supplied

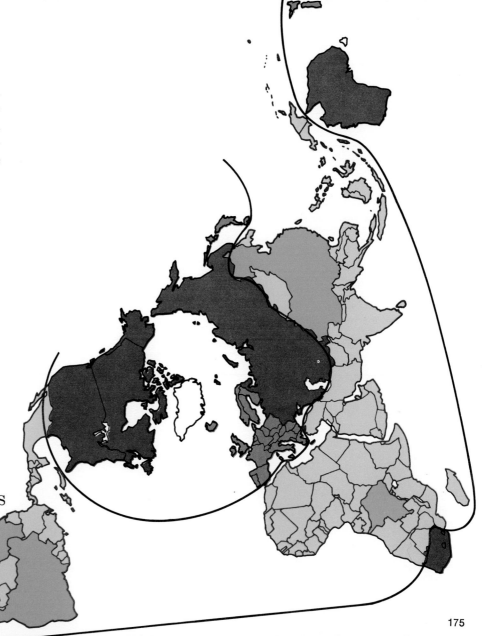

Minerals and the Major Producers (% of world total)

	Iron		Cobalt		Chrome		Manganese		Molybdenum		Nickel		Tungsten		Vanadium		Bauxite	
P: production R: reserves	P	R	P	R	P	R	P	R	P	R	P	R	P	R	P	R	P	R
Major producers																		
South Africa					34	67	23	41							30.5	42		
Australia	10.5	11	6	11			7.5	6					6				28.5	21
Canada	5.5	8.5							10.5	7	26.5	15		15				
United States	8.5	5							65	54			6	8	13,5			
USSR/Eastern countries	28	30	13	13	37.5	?	39	44	11	9	24	14	17	7.5	30.5	39	5.5	
Important Producers																		
Bolivia													6					
Brazil	11.5	20					8										5	11
Chile									11	25								
China	8						6						26	47	14	13		
India	5.5	5.5					6											
Mexico																		
Peru																		
Zaire			51	38														
Other producers																		
Algeria																		
Botswana			5															
Cuba											5.5	5.5						
Spain																		
Finland															7			
Gabon							6											
Guinea																	16	26
Indonesia																		
Jamaica																	14	9
Malaysia																		
Norway																		
N. Caledonia (Fr.)											7	25						
Philippines					6													
Thailand																		
Zambia			9	12														
Zimbabwe					6	30												
Total (%) prod. and world reserves	77.5	80	84	74	83,5	97	95.5	91	97,5	95	63	64.5	61	77,5	81,5	94	69	67

Notes :
- Only production and reserves over 5% of the world total are counted.
- ? Indicates probable important production or reserves (lack of data, principally USSR and China).

P: production / R: reserves	Copper		Tin		Lead		Zinc		Silver		Platinum		Antimony		Mercury		Titanium	
	P	R	P	R	P	R	P	R	P	R	P	R	P	R	P	R	R	R
Major producers																		
South Africa											45	81	16	7			15	5
Australia					11.5	14	8.5	10									7	7
Canada	9	7			10.5	13	17	26	10.5	19	5						25	
United States	19.5	18			13	27		20	10.5	21					14.5	7	6	
USSR/Eastern countries	12	7	15	10	17	11	17	9	14.5	30	48	16	12	20	32.5	11	?	?
Important Producers																		
Bolivia			10										23.5	8				
Brazil																		74
Chile	13.5	19																
China				15									18	39		?	20	
India																		
Mexico							5.5		14	13								
Peru		6			6		8.5		13									
Zaire	6																	
Other producers																		
Algeria															15	8		
Botswana																		
Cuba																		
Spain															15	33		
Finland																		
Gabon																		
Guinea																		
Indonesia			13	16														
Jamaica																		
Malaysia			25	12													21	
Norway																		
N. Caledonia (Fr.)																		
Philippines																		
Thailand			14	12														
Zambia	7.5	6																
Zimbabwe																		
Total (%) prod. and world reserves	67.5	63	77	65	58	65	56.5	65	62.5	83	98	97	69.5	74	77	59	94	86

(Titanium R columns: Ilmenite | Rutile)

• A number of useful minerals (boron, lithium, magnesium, niobium, strontium, tantalum, thorium) have been left out of this table for lack of data
• Minerals used for fertilizer (potassium, phosphates, etc.) are not included here.
• Titanium: derived primarily from two ores, ilmenite and rutile (incomplete data).

Source: U.S. Bureau of Mines.

Energy and Mineral Resources in the Oceans

Offshore mineral deposits such as deep-sea oil fields are always associated with the continental shelf. Gradually, technological developments are making it possible to operate at greater and greater depths. The number of fields being exploited has increased enormously in the last fifteen years (from 70 in 1965 to over 400 in 1980).

Since 1950, oceanographic research has revealed the mineral riches of the deep sea, which are in the form of metalliferous mud and polymetallic nodules. The importance of these deposits is promising for the not too distant future.

Oil wells

Minerals being exploited ★

Oil and natural gas exploitation ●

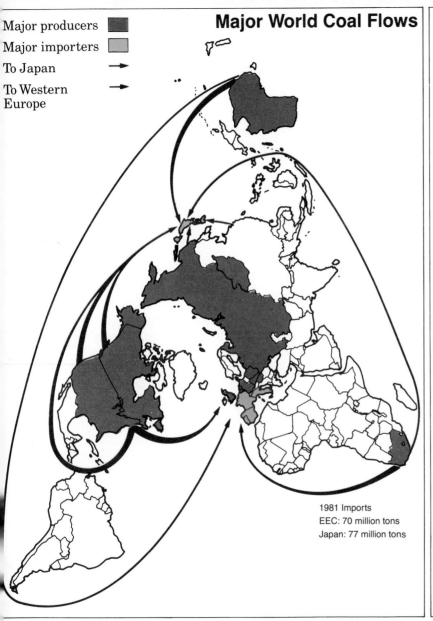

Major World Coal Flows

Major producers

Major importers

To Japan

To Western Europe

1981 Imports
EEC: 70 million tons
Japan: 77 million tons

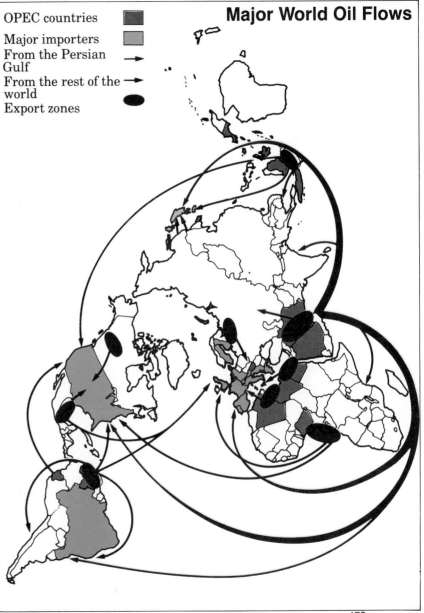

Major World Oil Flows

OPEC countries

Major importers

From the Persian Gulf

From the rest of the world

Export zones

Oil

Since the oil shocks of the 1970s, the demand for oil has slowed down considerably. Western consumption has been reduced by 25% in ten years. Production is now stagnant, at around 3 billion tons. The high-population Third World producers (Nigeria, Indonesia, Mexico, etc.) are suffering a severe recession after a few fat years. Saudi Arabia remains both the vital producer and the market regulator by virtue of its financial power.

World Production

1955	7.72 million tons
1970	2.28 million tons
1978	3.09 million tons
1981	3.05 million tons

Known Reserves
(90 billion tons)

Middle East	54%
Saudi Arabia	25%
Kuwait	10%
Iran	8%
Iraq	5%
USSR	11%
Mexico	9%
USA	4%

Producers

USSR	21%
Total Mid. East*	32%
Saudi Arabia	17%
Iraq	1%
USA	16%
Venezuela	4%
Mexico	4%
China	3%
Nigeria	2%

*Includes Iran, Kuwait, Bahrain, United Arab Emirates, Oman.

Coal

Production

USA	25%
USSR	19%
China*	24%
Poland	5%
UK	4-5%

Reserves

USA	31%
USSR	19%
China	25%
UK	9%
W. Germany	4%

*Combined coal and lignite.

Lignite

World Production

GDR	26%
USSR	16%
W. Germany	13.5%
Czechoslovakia	10%

Reserves

USSR	65%
USA	7%
E. Germany	10%
W. Germany	16%

Since 1973, the rise in oil prices has given a boost to coal production, which had been growing slowly since 1955.

1955—1.6 billion tons
1973—2.3 billion tons
1980—3.3 billion tons

Reserves are enormous, probably 80% of all fossil energy reserves. The appearance of new techniques making possible the burning of coal in the deposits themselves seems likely to open up new possibilities.

Members of OPEC (Organization of Petroleum Exporting Countries). Thirteen countries. (Arab countries have formed OAPEC, Organization of Arab Petroleum Exporting Countries.) Initial members included: Iraq, Iran, Kuwait, Libya, Saudi Arabia, and Venezuela (1960). Others followed, including Qatar (1961), Indonesia (1962), Abu Dhabi and the United Arab Emirates (1967), Algeria (1969), Nigeria (1971), Ecuador (1973), Gabon (1975). The 13 states supply more than 50% of the world production of unrefined oil.

Electricity*

Hydroelectricity

USA	20%
Canada	16%
USSR	10%
Brazil	
Japan	
Norway	} 4-5% each
France	
Sweden	

Nuclear-powered Electricity**

	1980	1990
USA	37.5%	28.5%
Japan	11.7%	7.7%
USSR	10.0%	17.5%
France	8.7%	12.8%
W. Germany	6.2%	5.4%
Canada	5.7%	3.2%

*Electricity furnished by power stations using combustible fossil fuels is not included.

**Despite disputes and resistance, the progress in establishing power stations throughout the world is spectacular.

Natural Gas

Production		Reserves	
USA	35%	USSR	40%
USSR	30%	Iran	18%
Netherlands	5%	USA	7%
Canada	3%	Algeria	4%
		W. Europe	5%

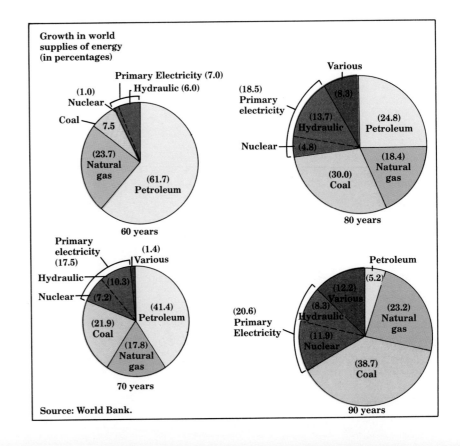

Growth in world supplies of energy (in percentages)

60 years
Primary Electricity (7.0)
Hydraulic (6.0)
(1.0) Nuclear
Coal 7.5
(23.7) Natural gas
(61.7) Petroleum

80 years
Various
(8.3)
(18.5) Primary electricity
(13.7) Hydraulic
Nuclear (4.8)
(24.8) Petroleum
(18.4) Natural gas
(30.0) Coal

70 years
Primary electricity (17.5)
(1.4) Various
Hydraulic (10.3)
Nuclear (7.2)
(41.4) Petroleum
(21.9) Coal
(17.8) Natural gas

90 years
Petroleum (5.2)
(12.2) Various
(8.3) Hydraulic
(20.6) Primary Electricity
(11.9) Nuclear
(23.2) Natural gas
(38.7) Coal

Source: World Bank.

Energy Consumption per Capita

(In tons of petroleum)

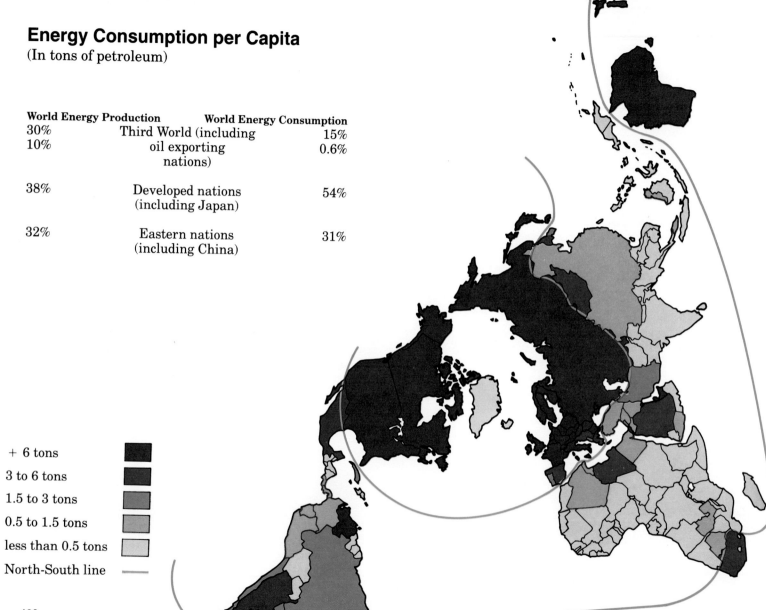

World Energy Production		World Energy Consumption
30%	Third World (including	15%
10%	oil exporting nations)	0.6%
38%	Developed nations (including Japan)	54%
32%	Eastern nations (including China)	31%

+ 6 tons

3 to 6 tons

1.5 to 3 tons

0.5 to 1.5 tons

less than 0.5 tons

North-South line

Technological Levels

High-tech industries employ new and very advanced technologies that depend heavily on research and are produced by only a very small number of industrialized countries, which are anxious to retain the virtual monopoly of them and so guarantee their supremacy. Such industries include aeronautics and aerospace, telecommunications and telematics, nuclear energy, computers, and bio-industry.

Data Processing (An example)

Computers in Service in the USA (% of world total)

1975	1980	2000
0.22 million	1.25 million	2.85 million
65%	65%	40%

World Computer Market (1981)

USA	73%
Japan	10%
EEC	16%

World Market for Electrical Components (1981)

USA	48%
Japan	27%
EEC	23%

High technology

Classic technology

Technological dependence

North-South lines

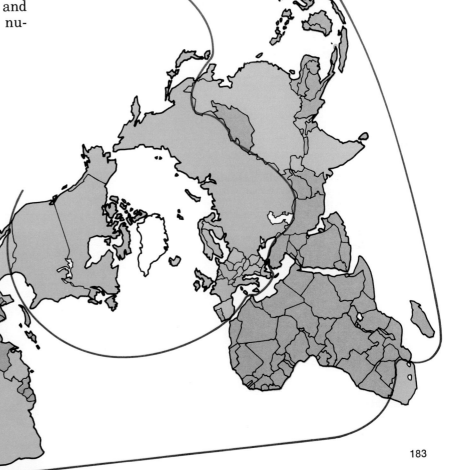

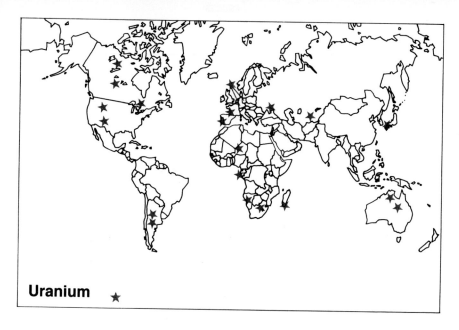

Uranium

Uranium

The states that are big producers and possessors of large reserves* are few and concentrated in North America and the southern hemisphere. There are no reliable figures for China, the USSR, and Eastern Europe.

Production

USA	33%
Canada	17%
S. Africa	13%
Namibia	9%
Niger	10%
France	6%
Australia	6%

Proved Reserves

Based on 5 million tons, except for the USSR, China, and Western countries that have reserves between 3.5 and 7 million tons.

USA	37%
Canada	19%
S. Africa	11%
Australia	7%
Sweden	6%
Niger	5%

Percentage Share of World Industrial Power 1980
(as a percentage of the combined value of mining and manufacturing output)

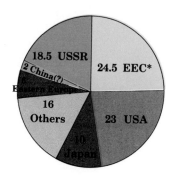

*EEC: FRG 9%, France 5.5%,
United Kingdom 4.5%.

---

*Thorium, occasionally used as a substitute for uranium, is abundant in India (30%), Canada (20%), and Brazil (6%).

World Grain Trade

**Major exporters
(% total exported)**

USA	54%
Canada	12%
Argentina	9%
France	7%
Australia	6%

Producers-exporters

Importers

Principal flows

Chronic malnutrition   *

Source: World Bank.

World Cereal Producers*		Major Cereals	
		Wheat	
USA	18%	USSR	19.0%
China	17%	USA	14.5%
USSR	10%	China	12.0%
India	9%		
France	3%	*Corn*	
Canada	3%	USA	46%
		China	13%
		Brazil	5%
Soya		*Rice*	
USA	63%	China	35%
Brazil	17%	India	20%
China	9%	Indonesia	8%

*Calculated on the basis of 8 cereals: wheat, rice, corn, barley, rye, oats, buckwheat, millet.

Fertilizers

Chemical fertilizers are now intensively used in countries with a modern agriculture to ensure high and regular yields, and also to combat soil deterioration.

The major fertilizer producers are the USA, the USSR, the EEC (more than half the total), and Canada.

World Imports of Cereals, by Groups of Countries, 1970 and 1980 (in percentages)

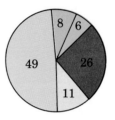

1970 = 109 million tons

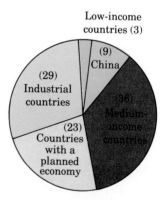

Low-income countries (3)

(9) China

(29) Industrial countries

(36) Medium-income countries

(23) Countries with a planned economy

1980 = 228 million tons

Source: World Bank.

Livestock

The real value of livestock is difficult to compare because of the enormous variety of species and modes of stock-rearing. High-productivity Danish cattle-rearing and the livestock of India, for example, have nothing in common.

Over the last two decades, factory farming of chickens and pigs has vastly increased in the developed countries, largely replacing traditional methods of husbandry.

Cattle		Sheep		Pigs	
India	15%	USSR	13%	China	38%
USSR	10%	Australia	12%	USSR	10%
USA	9%	China	9%	USA	8.5%
Brazil	7.5%	N. Zealand	6%	Brazil	5%
China	5%				
Argentina	5%				

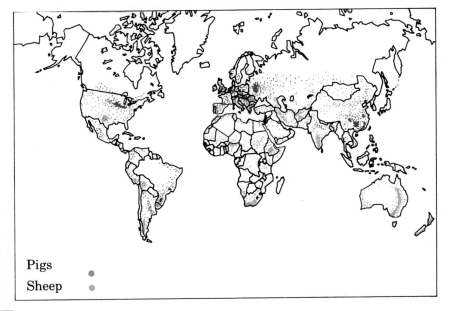

Pigs

Sheep

Fishing

Fishing supplies about 10% of humanity's protein requirements. In addition to food, industrial fishing produces a number of products used in industry: flours, oils, grease, etc.

If maritime states extend the limits of their territorial waters to 200 nautical miles, the industrial and deep-sea fishing activities of other states are likely to be severely handicapped.

Japan	14.6%
USSR	12.5%
China	5.8%
USA	5%
Chile	4%
Peru	4%
Norway	4%

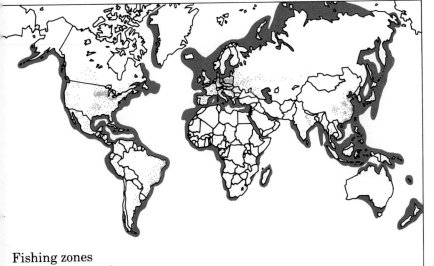

Fishing zones

Cattle

POPULATION FACTORS

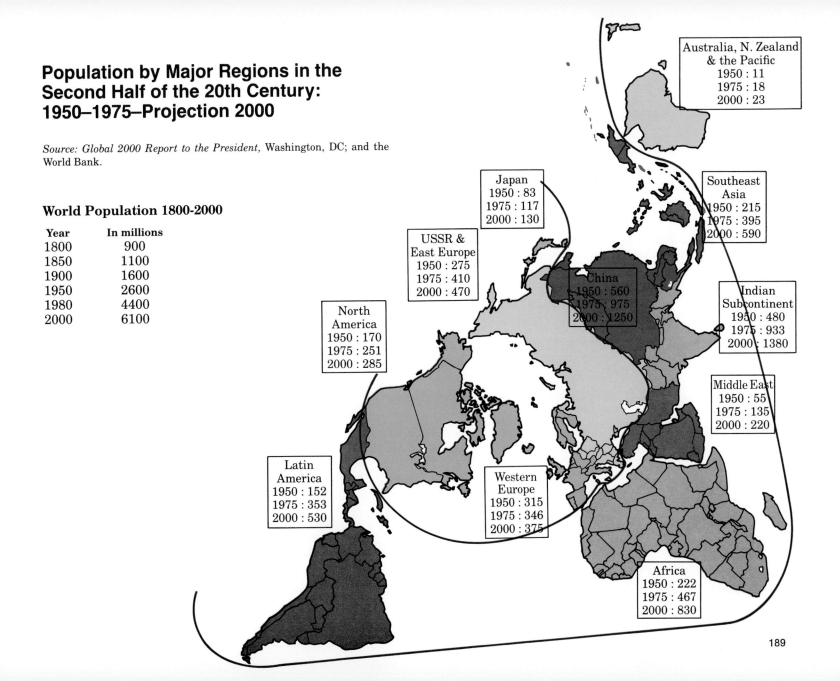

Population by Major Regions in the Second Half of the 20th Century: 1950–1975–Projection 2000

Source: Global 2000 Report to the President, Washington, DC; and the World Bank.

World Population 1800-2000

Year	In millions
1800	900
1850	1100
1900	1600
1950	2600
1980	4400
2000	6100

Australia, N. Zealand & the Pacific
1950 : 11
1975 : 18
2000 : 23

Japan
1950 : 83
1975 : 117
2000 : 130

Southeast Asia
1950 : 215
1975 : 395
2000 : 590

USSR & East Europe
1950 : 275
1975 : 410
2000 : 470

China
1950 : 560
1975 : 975
2000 : 1250

Indian Subcontinent
1950 : 480
1975 : 933
2000 : 1380

North America
1950 : 170
1975 : 251
2000 : 285

Middle East
1950 : 55
1975 : 135
2000 : 220

Latin America
1950 : 152
1975 : 353
2000 : 530

Western Europe
1950 : 315
1975 : 346
2000 : 375

Africa
1950 : 222
1975 : 467
2000 : 830

Most Populous States: Evolution of Their Population, 1930 to 2000 (est.)

	1930	1950	1980	2000
China.........	430	540	970	1240
India	335	370	673	975
USSR........	160	180	265	315
USA.........	120	155	227	260
Indonesia.....	60	77	146	220
Brazil........	40	53	119	177
Bangladesh ...		40	92	148
Pakistan		35	82	141
Nigeria	19	28	85	161
Japan........	64	84	117	130
Mexico.......	16	26	67	110
Vietnam......	22	25	54	88
Philippines ...	12	20	48	75
Thailand	11	19	46	68
Turkey.......	11	21	45	70
Iran	14	20	38	64
Egypt........	14	21	40	61
Italy.........	41	47	57	61

Evolution of rhythm of world increase (mean annual rate)

1800-1850: 0.55%
1850-1900: 0.57%
1900-1950: 0.83%
1950-1980: 1.87%
1980-2000: 1.80% est.

AGE AND SEX OF THE WORLD POPULATION, 1975 AND 2000

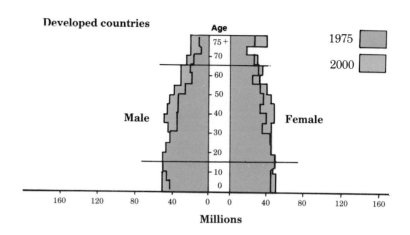

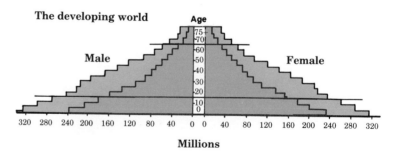

Source: Global 2000 Report to the President, Washington, DC, 1980.

World Urbanization

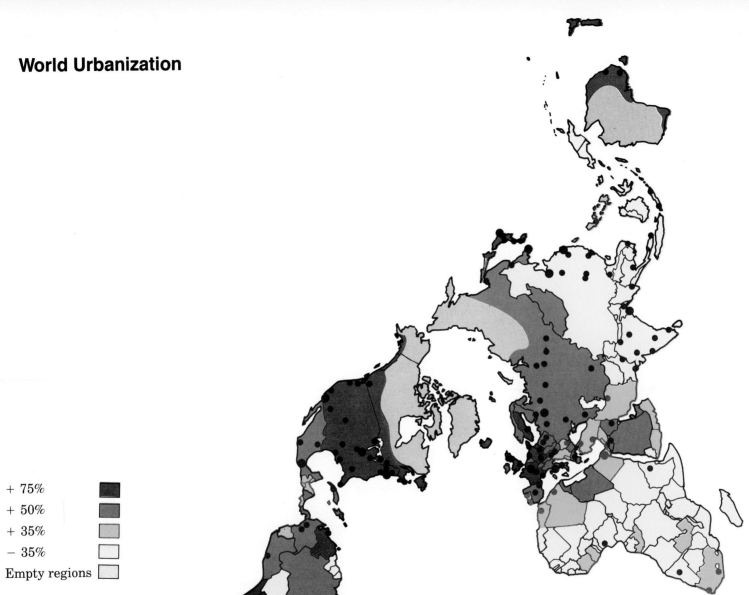

+ 75%

+ 50%

+ 35%

− 35%

Empty regions

Depopulated areas are shown
for the USSR, Canada,
Australia.

Urbanization

The Major Cities or Concentrations of the World

(In millions of inhabitants; the powerful industrial cities are underlined.)

1900		1950		2000 (est.)	
London	6.4	New York	12.3	Mexico	31
New York	4.2	London	10.4	S. Paulo	25.8
Paris	3.9	Rhine-Ruhr	6.9	Shanghai	23.7
Berlin	2.4	Tokyo	6.7	Tokyo-Yokohama	23.1
Chicago	1.7	Shanghai	5.8	New York	22.4
Vienna	1.6	Paris	5.5	Peking	20.9
Tokyo	1.4	Buenos Aires	5.3	Rio	19
St Petersburg	1.4	Chicago	4.9	Bombay	16.8
Philadelphia	1.4	Moscow	4.8	Calcutta	16.4
Manchester	1.2	Calcutta	4.6	Jakarta	15.7
Birmingham	1.2	Los Angeles	4	Los Angeles	13.9
Moscow	1.2	Osaka	3.8	Seoul	13.7
Peking	1.1	Milan	3.6	Cairo	12.9
Calcutta	1	Bombay	3	Madras	12.7
Boston	1	Mexico	3	Buenos Aires	12.1
Glasgow	1	Philadelphia	4	Karachi	11.6
Liverpool	0.98	Rio	2.9	Delhi	11.5
Osaka	0.95	Detroit	2.8	Manila	11.4
Constantinople	0.92	Naples	2.6	Teheran	11.1
Hamburg	0.9	Leningrad	2.5	Baghdad	11

URBAN POPULATION

Source: World Bank.

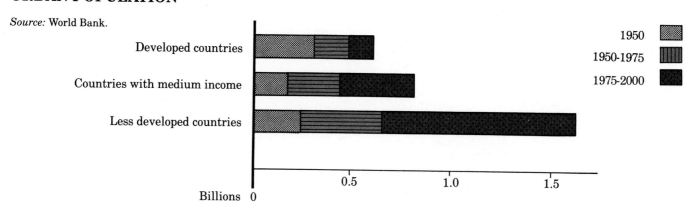

Developed countries

Countries with medium income

Less developed countries

1950

1950-1975

1975-2000

Billions 0   0.5   1.0   1.5

NORTH–SOUTH

As capitalism introduced modernizing elements abroad, it dislocated the traditional economies of the dominated countries. These distortions lie at the origin of what is called underdevelopment. In fact, the Third World covers very diverse realities, although it is possible to list a number of common traits: preponderance of the agricultural sector, high birth rate, very marked economic and social inequalities. Over the last two decades, the Third World has seen growing differentiations caused by both the level of local productive forces and/or the significance of mineral resources, particularly oil and natural gas. Outside a handful of countries that have seen high growth rates, the majority of Third World countries have made little progress and have generally stagnated or even regressed. In a majority of countries, agriculture, which still employs the bulk of the population, cannot satisfy local needs. Deterioration in the terms of trade that has marked the three decades following World War II continues to weigh heavily. Only the OPEC countries, producers of oil and natural gas, were able to modify the laws of the market during the decade 1973 to 1983.

Population growth, which has since the beginning of the century been consistently high in the Third World, continues at a rate approaching 3%. World population, estimated at 4 billion ten years ago, will pass 6 billion by the end of the century. At that date, peoples of the Third World will make up 80% of the globe's population, whereas they were only 65% in 1970. The living conditions of the 2.5 billion people who live in the Third World have scarcely improved in twenty years, while 880 million of them are considered to be living in conditions well below subsistence. The reforms proposed by international organizations concerned with development run up against both the unequal North–South system and the political impediments constituted by the ruling strata in the Third World. Up to now, the countries that have recorded notable growth are the oil-producing countries, especially those with small populations, like Kuwait, Saudi Arabia, the United Arab Emirates, and Libya; a handful of countries, such as Brazil or Gabon (underpopulated) that have large mineral resources; and a series of Asian countries where the quality of manpower and business abilities are of a high order: Taiwan, Singapore, South Korea, Thailand, etc.

Today, there are three categories of countries in the Third World: those few that have a high or relatively high growth rate, some of which already have a significant industrial infrastructure; a sizable proportion of average countries that are developing despite a high population growth; and about fifty classified as less developed countries (LDCs) whose situation is tragic. The current crisis is hitting the Third World very hard. Africa seems particularly threatened in every respect: stagnation, or even regression of growth, galloping population growth, corruption, and the ineffectiveness of the vast majority of ruling strata. In Andean America (Ecuador, Peru, Bolivia, etc.), the situation—and the prospects—are scarcely any better. In Southeast Asia, alongside not insignificant growth and dynamic industrial sectors, there remain, especially in the Indian subcontinent (Bangladesh, India, Pakistan), enormous areas of poverty and malnutrition.

Evolution of GNP by Main World Regions
1975—Projection 2000
(In dollars per capita)

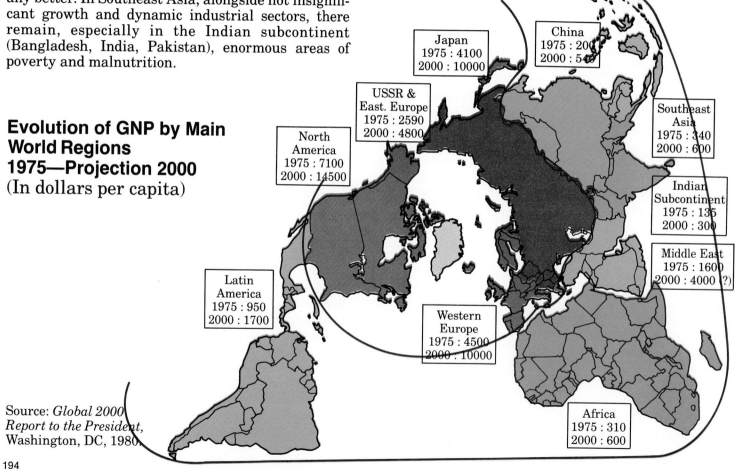

Australia
N. Zealand
1975 : 4900
2000 : 10000

Japan
1975 : 4100
2000 : 10000

China
1975 : 200
2000 : 540

USSR &
East. Europe
1975 : 2590
2000 : 4800

North
America
1975 : 7100
2000 : 14500

Southeast
Asia
1975 : 340
2000 : 600

Indian
Subcontinent
1975 : 135
2000 : 300

Middle East
1975 : 1600
2000 : 4000 (?)

Latin
America
1975 : 950
2000 : 1700

Western
Europe
1975 : 4500
2000 : 10000

Africa
1975 : 310
2000 : 600

Source: *Global 2000 Report to the President,* Washington, DC, 1980.

Income
Health
Education
1950-1980

Income

GNP per person (1980 dollars)	1950	1960	1980
Industrial countries	4,130	5,580	10,660
Medium-income countries	640	820	1,580
Low-income countries	170	180	250

Mean annual increase (%)	1950-1960	1960-1980
Industrial countries	3.1	3.3
Medium-income countries	2.5	3.3
Low-income countries	0.6	1.7

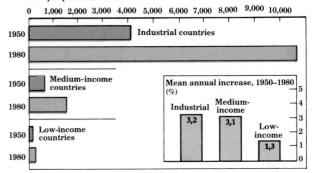

Health

Life expectancy at birth (years)

	1950	1960	1979	Increase 1950-79
Industrial countries	67	70	74	7
Medium-income countries	48	53	61	13
Low-income countries	37	42	51	14
Planned economies	60	68	72	12

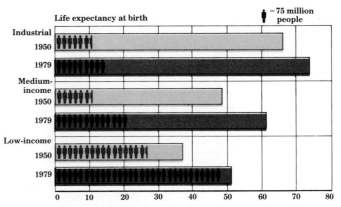

Life expectancy at birth

= 75 million people

Education

Adult literacy rate (%)

	1950	1960	1976
Industrial countries	95	97	99
Medium-income countries	48	53	72
Low-income countries	22	28	39
Planned economies	97	97	99

Note: China is not included.

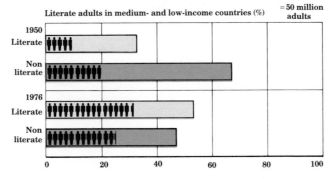

Literate adults in medium- and low-income countries (%)

= 50 million adults

Source: World Bank.

Evolution of Food Consumption
(1970–1980)

According to the World Bank, 800 million people in
the Third World live below the threshold of absolute
poverty and suffer from undernourishment or grave
malnutrition, especially in Asia and subtropical Africa.

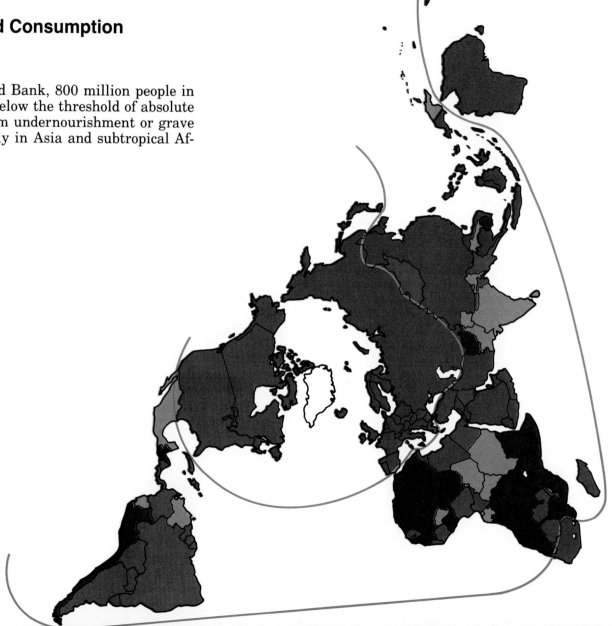

Increase

Stagnation

Decrease (0-10%)

Decrease (+ 10%)

North-South line ⎯

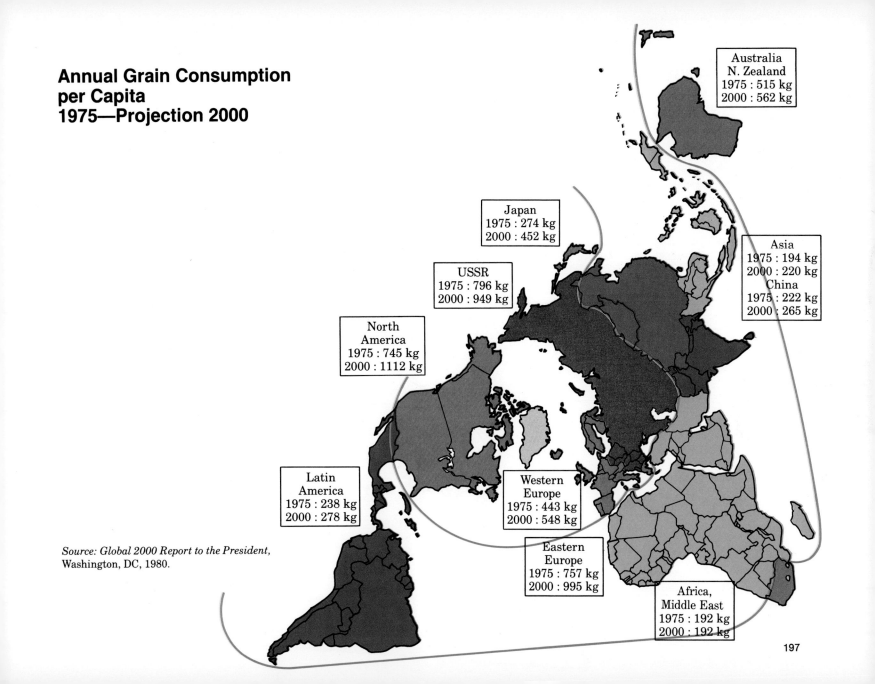

Annual Grain Consumption per Capita 1975—Projection 2000

Australia
N. Zealand
1975 : 515 kg
2000 : 562 kg

Japan
1975 : 274 kg
2000 : 452 kg

Asia
1975 : 194 kg
2000 : 220 kg
China
1975 : 222 kg
2000 : 265 kg

USSR
1975 : 796 kg
2000 : 949 kg

North
America
1975 : 745 kg
2000 : 1112 kg

Latin
America
1975 : 238 kg
2000 : 278 kg

Western
Europe
1975 : 443 kg
2000 : 548 kg

Eastern
Europe
1975 : 757 kg
2000 : 995 kg

Africa,
Middle East
1975 : 192 kg
2000 : 192 kg

Source: Global 2000 Report to the President,
Washington, DC, 1980.

Major World Industrial Regions

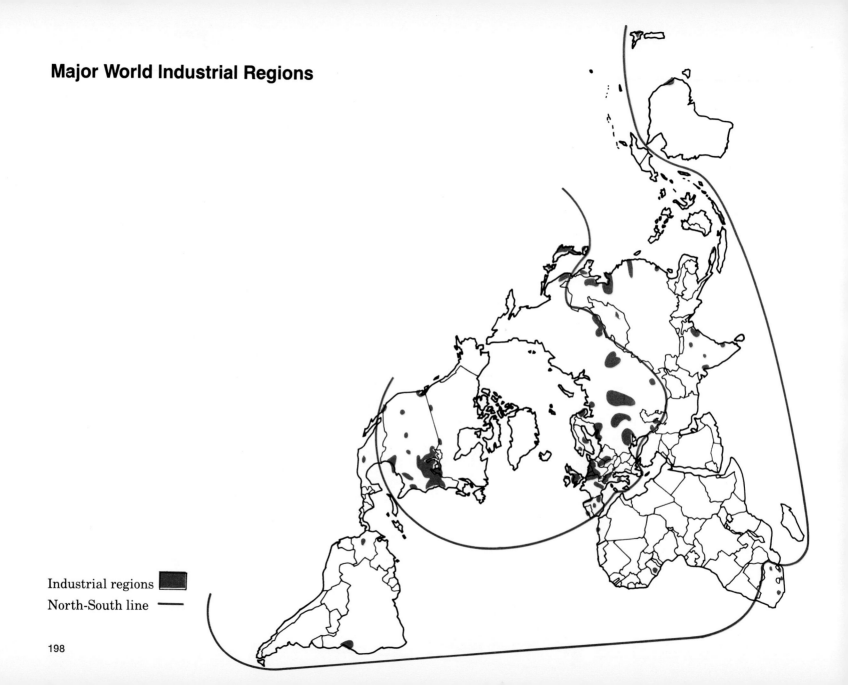

Industrial regions ▬
North-South line ▬▬▬

The Third World: East–West Struggle

What is the interim balance sheet of East–West competition in the Third World?

The period immediately after World War II was favorable to the Communist forces in Asia: in China totally, in Korea partly. After a long rearguard action waged by the French and the Americans, the upholders of "Marxism–Leninism" won a major success in Indochina, of which Vietnam was the major beneficiary.

Conversely, in Indonesia in 1965, the largest Communist party in Asia was decimated. Over time, the economic situation of the regional allies (Taiwan, Singapore, South Korea, Thailand, Malaysia) of the United States improved.

The recognition of China in 1971 (after a decade of Sino-Soviet conflict), as a result of the Nixon-Kissinger initiative, enabled the United States to engage in a three-sided game advantageous to it.

In the Middle East, compared with the Baghdad Pact (Turkey, Iraq, Iran, Pakistan) period, the West has suffered serious setbacks. The fall of the Shah in Iran was a major setback for the USA. But after the long alliance during the Nasser period, the breach precipitated by Egypt (1972) was just as serious for the USSR, whose Arab policy has run into difficulties (Iraq) and is only very partially effective (Syria, PLO) in a situation where, up till now, the United States continues to be in the position of arbiter. Israel, a close ally of the United States, remains the major military power in the region, while Saudi Arabia, whose stability is vital for the West, has played a moderating role among the oil-producing countries.

Russian occupation of Afghanistan, conversely, weakened the United States ally Pakistan, whereas India has succeeded since Nehru in making the best use of the delicate notion of nonalignment.

The Western preserve that Africa had been up to 1975, despite various Soviet attempts (from Guinea 1959, to Sudan and Somalia in the early 1970s) is now an integral part of the East-West confrontation. In Angola (1975) and Ethiopia (1977), the presence of Cuban troops added an altogether new dimension to the Soviet presence. Mozambique can be included among the allies of the USSR. Other African countries, sometimes classified as allies of the USSR, seem to be in a much more ambiguous situation: Congo, Guinea-Bissau, Madagascar, etc. On the continent as a whole, the West continues to be predominant.

In Latin America, the United States (except during the Carter administration) has essentially maintained its role as policeman: Directly or indirectly, the policy of the "big stick" has been used in Guatemala (1954), without success in Cuba (1961), in the Dominican Republic (1965), in the training of counterinsurgency forces throughout the continent to deal with the guerrilla movements of the 1960s, as well as in Brazil (1964) and Chile (1973).

The new factor, since the 1979 victory of the Sandinistas in Nicaragua (made easier by American neutrality), is the increase in the number of large-scale armed struggles in Central America (El Salvador, Guatemala). These guerrilla wars, born of local economic situations, enable the USSR and Cuba to challenge Washington in its own backyard, the Caribbean basin, which is vital for the geostrategic security of the United States.

Overall, it is less a U.S. retreat (in any case only relative and altogether normal, given the fact that the United States held a virtually worldwide hegemony in a situation of complete military superiority) than the transformation of the USSR from a regional power into a truly world power, that is the major change of the last three decades.

In this new context, it goes without saying that the world crisis, felt particularly in the Third World and aggravated by the fecklessness of the ruling strata in many countries, will make the confrontations sharper.

THE MOVEMENT OF THE NONALIGNED
Created in 1961 at the Belgrade Conference.
Latin America: Argentina, Bolivia, Peru, Ecuador, Panama, Nicaragua, Cuba, Jamaica, Guyana (ex-British), Surinam, Trinidad and Tobago, Grenada, Barbados and St. Lucia.
Africa: All states except Namibia, South Africa, and Western Sahara (ex-Spanish); the Cape Verde Islands, São Tomé and Principe, Mauritius, the Comoros, and the Seychelles.
Europe: Yugoslavia, Malta, and Cyprus.
Middle East: All the Arab states, Iran, Afghanistan.
Indian subcontinent: Pakistan, India, Nepal, Bhutan, Bangladesh, Sri Lanka, the Maldives.
East Asia: Laos, Vietnam, Cambodia, Malaysia, Indonesia, Singapore, North Korea.
Le Groupe des 77: An organization for economic order concerned with North-South relations; includes 120 members (1980) including all the countries of the Third World except China, Namibia, Western Sahara (ex-Spanish), and Turkey. Also members: Yugoslavia, Romania.

Arms Sales to Third World Countries

The leading arms exporters are: the United States (37%), the USSR (30%), France (8%), and Great Britain and West Germany (5% each).

The major arms importers are in the Middle East (40%). If members of NATO (12%) and the Warsaw Pact (12%) are excluded, the other importers are Africa (17%), Asia (11%), Latin America (5%). In order of quantity, the main purchasers are Saudi Arabia, Jordan, Iraq, Syria, Libya, South Korea, India, Israel, Vietnam, Morocco, and Ethiopia. These countries (plus Iran*) buy almost a third of the total arms sold in Asia, Africa, and Latin America.

———
*It is difficult to know what place Iran occupies at present.

Indebtedness: A Factor Aggravating the Crisis

Total Public Debt, Third World Countries, 1982: $575 billion* (1973: $98 billion)

Most indebted states:
(Public debt and pledges in $ billions)

Brazil	43.8	Algeria	14.4
Mexico	42.7	Egypt	13.8
Poland	27(?)	Turkey	13.8
S. Korea	20.0	Israel	13.8
India	18.5	Venezuela	11.3
Indonesia	15.5	Argentina	10.0

Eastern European Countries Public Debt: $80 billion (1973: $8 billion)

*Interest payments—$31 billion in 1980—doubled between 1978 and 1980.

The changes that have occurred in the debt structure of developing countries contrast with the stability of its distribution. Since 1973, it is the same twelve major borrowers who owe almost two-thirds of the debt and debt servicing, each of them having at present a debt of over 12 billion dollars. Five of these debtors—Algeria, Egypt, Indonesia, Mexico and Venezuela—are oil-exporting countries. The other seven—Argentina, Brazil, India, Israel, Korea, Turkey, and Yugoslavia—are major exporters of manufactures.

It is middle-income countries that accounted for most of the debt of developing countries (349 billion dollars out of a total of 410 billion dollars) at the end of 1980. About 70% of their debt had been incurred from private sources.

Conversely, at the end of 1980, 87% of the debt of low-income countries had been contracted from public sources; out of net transfers of 5 billion dollars made to these countries during 1980, public sources of financing supplied 88%. Whereas the total amount of flows to these countries increased by 55% in 1980 to reach 13.5 billion dollars, those made by multilateral institutions almost doubled (and reached 6.4 billion dollars).

Source: World Bank.

External Public Debt
(In % of GNP for states of each group)

Poor countries: 19.2% (1970: 15.6%)

Middle-income countries:
Petroleum importers: 15.4% (1970: 10.7%)
Petroleum exporters: 21.3% (1970: 14%)

Eastern countries: no data

Yugoslavia: 6.6% (1970: 8.8%)

80% of the debt of the private banking system to countries in the process of development is held by 20 developed countries, and 50% by only five among those: the United States, Canada, the United Kingdom, France, and West Germany.

Remittances of Funds Received by Labor-Exporting Countries

Major Countries	Amount in $ Billions	Remittance of Funds as % of Exports of Merchandise
Yugoslavia	4.8	70
Turkey	2.0	71
Italy	3.2	2
Algeria	0.5	3
Spain	1.2	6
Tunisia	0.3	17
India	1.2	18
Pakistan	2.0	80
Greece	1.0	26
Morocco	1.0	44
Portugal	2.9	64
Jordan	0.8	138
Egypt	2.6	220
S. Yemen	0.3	715
Yemen	1.3	995

Source: OECD, 1980.

THE MILITARY BALANCE

The Military Balance

Compiling a balance sheet using tables listing the armed might of the protagonists is only one approach.*

This statistical listing, though very useful, is only one admittedly vital factor in a balance in which sociopolitical factors, attitudes, and will remain of fundamental importance. Traditionally, the listings are measured by experts, but strategies are won with peoples and leaders.

For thirty-five years, generalized war has been avoided through the balance of terror. Conflicts have taken place in the framework of indirect strategies, in local conventional wars, or still more often in guerrilla movements or crisis management without overt conflict (Berlin, Cuba).

During the last twenty years, since the Cuban missile crisis (1962), the USSR has greatly increased its military might in the air, on the sea, and in the nuclear domain.

In the present state of affairs, while nuclear war seems highly improbable, nations must act as if it were possible (at least by accident)—hence the need, in the arms race, to be constantly improving their panoply of weapons.

If, in short, it is agreed that nuclear war, launched in cold blood, is not at present (but for how long?) conceivable, each side has to respond to the expectations of the other. In this respect, today as yesterday, strategy rests on a "dialectic of uncertainties" (Lucien Poirier).

But strategy is not limited to war—far from it. It aims above all at producing psychological effects on peoples.

In connection with a possible resumption of discussions on arms control and reduction (SALT I, 1972), everything points to the conclusion that the installation of SS-20s by the Soviets (1977) was a coercive strategy, intended, through the prospect of the threat of a nuclear war, to lead to the neutralization of Western Europe vis-à-vis the United States—what in the jargon is called political uncoupling.

The SS-20† is not a weapon of deterrence but of pressure on Western Europe and especially on West Germany, its Achilles' heel. In this sense, the debate on the installation of Pershing 2s is first of all a problem of political will for Europeans.

---

*In addition, figures are quickly out of date. See *The Military Balance* by the International Institute of Strategic Studies, London, which annually provides figures on the world military balance, nation by nation.

---

†Each SS-20 carries three independently targeted nuclear warheads, with very great firepower. Carrying 150 kilotons in each nuclear warhead, the SS-20 has a range of 2,700 nautical miles. At present, the USSR has 300 SS-20s. Experts estimate that 150 SS-20s could destroy the whole of NATO's defense system in Europe.

Strategic Nuclear Arsenals

	Atlantic Alliance	Warsaw Pact
Nuclear attack submarines	90	99
Combat vessels, tonnage	3.385*	2.685*
Aircraft carriers	18	4
Vessels of over 2,000 tons	348	139
Amphibious vessels, tonnage	759*	141*
Navy planes	2,600	750

*In millions of tons.

	USA	USSR
Intercontinental missiles	1,052	1,398
Nuclear warheads	2,152	5,230
Strategic ballistic missiles, submarine launched	520 On board 36 nuclear-powered submarines	969 On board 80 nuclear-powered submarines
Nuclear warheads	4,768	1,752
Continental strategic missiles, surface to surface	*	740 Of which 80% are west of the Urals
Strategic bombers	376 Of which 316 are long-range	150 Of which 105 are long-range

Source: Military Balance, 1982–83.

*Pershing 2s and ground-launched cruise missiles are, in principle, to be installed in West Germany, Great Britain, Italy, Belgium, and the Netherlands.

France has 80 submarine-launched strategic missiles (range 3,000 km, with one nuclear warhead per missile, aboard 5 nuclear-powered submarines). It also has 18 intermediate-range strategic missiles (3,500 km) and 36 strategic bombers.

Great Britain has 64 submarine-launched strategic missiles (range 4,000 km, with three thermonuclear warheads per missile, aboard 4 nuclear-powered submarines).

As far as tactical nuclear forces are concerned, the Atlantic Alliance has 348 surface-to-surface missiles against 1,980 for the Warsaw Pact countries, and 603 airplanes against 2,650.

The comparisons reflect only very imperfectly the actual forces, even numerically, as far as air and land potentials are concerned. The American and Soviet tank and airborne divisions have quite different compositions. In fact, no one doubts the superiority of the Warsaw Pact's conventional forces in the European theater.

From the point of view of naval potential, the superiority of the Atlantic Alliance is beyond doubt, control of the seas being vital to the West.

Deployment of United States Naval Forces

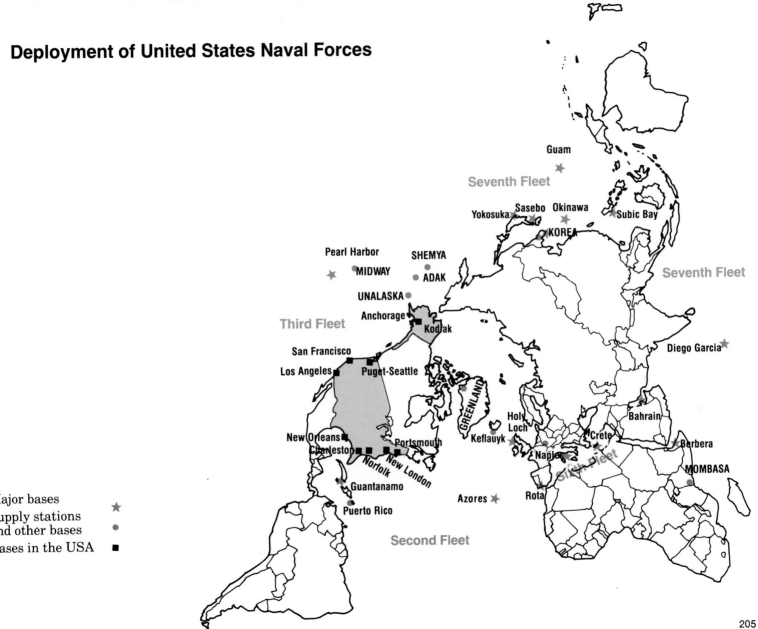

Guam

Seventh Fleet

Yokosuka Sasebo Okinawa Subic Bay

KOREA

Seventh Fleet

Pearl Harbor SHEMYA

MIDWAY ADAK

UNALASKA

Anchorage

Third Fleet Kodiak

Diego Garcia

San Francisco

Los Angeles Puget-Seattle

GREENLAND

Bahrain

Holy Loch

Keflauyk Crete

New Orleans Portsmouth Berbera

Charleston Naples

Norfolk New London Sixth Fleet MOMBASA

Guantanamo Rota

Azores

Puerto Rico Second Fleet

Major bases ★

Supply stations and other bases ●

Bases in the USA ■

205

Soviet World Naval Deployment and Supply Stations

This deployment is in effect despite seasonally frozen seas:
- toward the Atlantic from the White Sea and the Kola Peninsula
- toward the Pacific from the Siberian coast (Vladivostok–Petropavlovsk)

The military ports in the Baltic and the Black Sea (semi-closed seas) have seen their role decline.

The Soviet naval presence is particularly large in the Indian Ocean.

Vietnam and Cuba provide particularly reliable supply stations.

Major supply stations ■
Other supply stations □
Soviet military ports ●
Allies of the USSR ▨

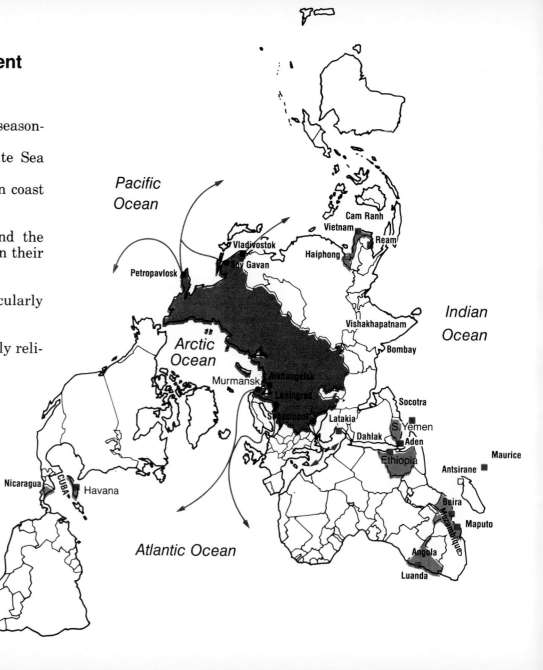

The Sea and Sea Power

The sea is the most widely used means of communications and trade, and the volume of trade has continuously increased over recent decades. The West—and Japan—depend heavily on freedom of navigation.

As a result of recent technological progress, the sea, a traditional source of wealth from fishing, has become an object of conflict, and disputes over the limits of sovereignty have increased. Offshore petroleum already accounts for a third of world production. But the very near prospect of the exploitation of significant ocean mineral resources is awakening appetites.

The Maritime Convention adopted in 1982 by most countries—but not by the United States, which is the best equipped technologically to exploit undersea riches—stipulates that territorial waters extend 12 miles and that states exercise sovereign rights (especially economic ones) up to a 200-mile limit.

During the 1970s, the rise of Soviet sea power, the result of two decades of effort, was one of the main factors changing the East-West balance, which had previously been almost exclusively limited to the land theater. From 1961 onward, the Soviet fleet progressively extended its presence, and by 1975 it was a worldwide force.

American naval superiority remains undeniable, but "the U.S. Navy must keep open the lines of communication without which the United States could not live, while the Soviet fleet does not have to protect vital lines of communication but only to cut those of its adversary" (Courtau-Bégarie). The new world order is largely being played out on the seas, with their resources and their bases.

WORLDWIDE DEPLOYMENT OF AMERICAN NAVAL FORCES

Second Fleet (Atlantic)
Major bases: *Norfolk,* Charleston, Jacksonville, New Orleans, Puerto Rico, Boston, New London, Brunswick (Portland)
• 72 submarines (with 31 missile launchers)
• 76 major combat vessels
Units of this fleet are at Guantanamo (Cuba), Bermuda, Keflavik (Iceland), Holy Loch (GB).

Third Fleet (Eastern Pacific)
Major bases: *Pearl Harbor* (Hawaii), San Francisco, San Diego, Whidbey I., Long Beach (Los Angeles), Adak (Alaska)
• 35 submarines (with 5 missile launchers)
• 47 major combat vessels

Sixth Fleet (Mediterranean)
Principal bases: Gaeta, Naples (Italy), Rota (Spain)
• 5 submarines
• 16 major combat vessels

Seventh Fleet (Western Pacific)*
Major bases: Yokosuka (Japan), Subic Bay (Philippines), Guam, Midway
• 8 submarines
• 23 major combat vessels

*A certain number of combat vessels are in the Indian Ocean and the Persian Gulf area (about 20 surface vessels)

Note: Below the names of the fleets appear their bases or base areas. The headquarters of the fleets are shown in italics.

WORLDWIDE DEPLOYMENT OF SOVIET NAVAL FORCES

North Fleet
Severomorsk, Kola, White Sea
- 135 submarines (with 45 missile launchers)
- 82 major combat vessels

Baltic Fleet
Baltiysk, Kronstadt, Riga, Tallin
- 22 submarines
- 42 major combat vessels

Black Sea Fleet
Sebastopol, Odessa, Poti
- 22 submarines
- 84 major combat vessels
(The Soviet squadron operating in the Mediterranean is included in this fleet.)

Pacific Fleet
Vladivostok, Petropavlosk, Sov. Gavan
- 80 submarines (with 24 missile launchers)
- 86 major combat vessels
(Elements of this fleet are stationed or operating in Vietnam—Camh Ranh, Danang—and in the Indian Ocean—Socotra, Aden, Kahlak, etc.)

Note: Below the names of the fleets appear their bases or base areas. The headquarters of the fleets are shown in italics.

AMERICAN MILITARY ABROAD*

Total	**469,711**
Land	*446,043*
Sea	*23,668*

Europe	328,577	Western Hemisphere	16,642
Far East & Pacific	124,492	At sea	2,138
At sea	21,530	Panama	9,616
Japan	47,269	Guantanamo	2,163
S. Korea	39,317	Other	2,725
Philippines	15,414		
Australia	636		
Other	326		

SOVIET MILITARY PRESENCE AND ALLIES, WORLDWIDE (1982)*

Strong presence Soviet bloc technicians or military advisers

Ethiopia	3,000-3,500	Cuba
Afghanistan	7,000-8,000	USSR
Algeria	2,000-3,000	USSR
Libya	2,000	USSR
Syria	2,000	USSR
South Yemen	1,500	USSR
Nicaragua	1,000	Cuba

SOVIET BLOC TROOPS ABROAD

	Country	Number	Since
USSR	Afghanistan	100,000	1979
Cuba	Angola	15,000-20,000	1975
	Ethiopia	7,000-9,000	1977
Vietnam	Cambodia	180,000	1978

*Source: U.S. Department of State.

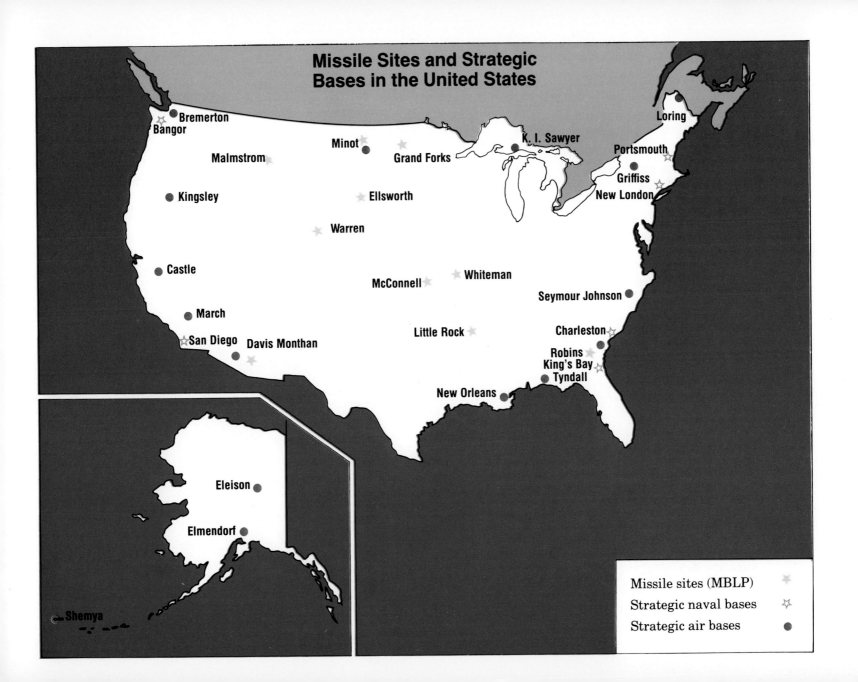

Missile Sites and Strategic Bases in the United States

Bremerton
Bangor
Minot
Grand Forks
K. I. Sawyer
Loring
Portsmouth
Malmstrom
Griffiss
New London
Kingsley
Ellsworth
Warren
Castle
McConnell
Whiteman
Seymour Johnson
March
Little Rock
Charleston
San Diego
Davis Monthan
Robins
King's Bay
Tyndall
New Orleans

Eleison
Elmendorf
Shemya

Missile sites (MBLP)

Strategic naval bases

Strategic air bases

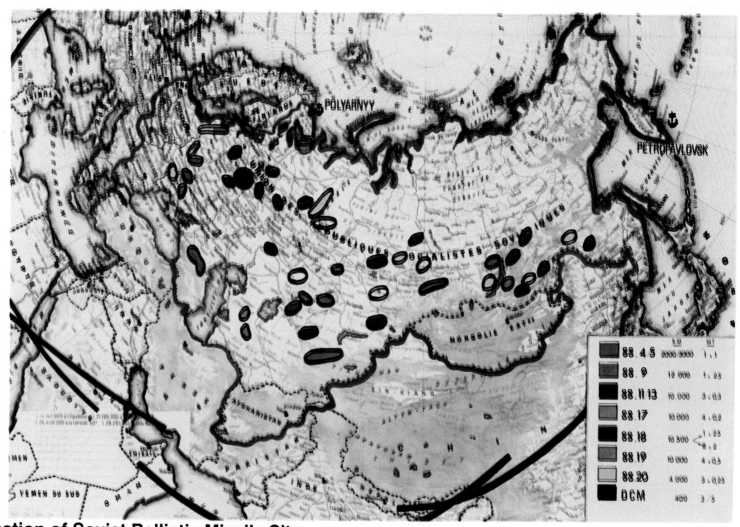

Location of Soviet Ballistic Missile Sites

Map supplied by General P. Gallois.

The USA under the Threat of Soviet Missiles

Positions occupied by
Soviet submarine
missile launchers able
to hit the territory of
the USA in 1970. No
major change in 1980.

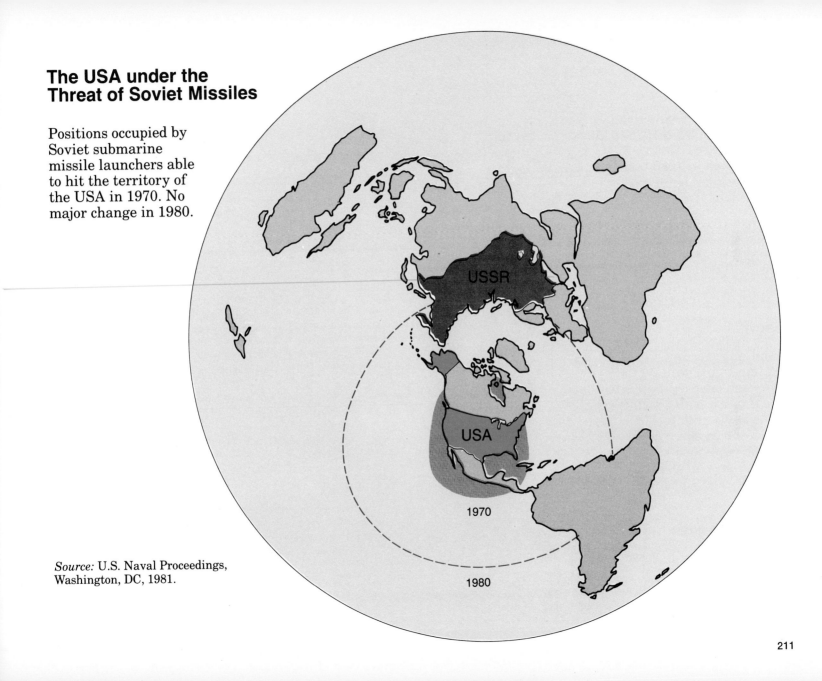

Source: U.S. Naval Proceedings,
Washington, DC, 1981.

The USSR under the Threat of American Missiles

Distance at which American missiles had to be in 1970 to hit Soviet territory.
Their position in 1980 to hit the same target (Trident I). Before 1990, Trident II will make it possible to extend the firing range farther.

Source: U.S. Naval Proceedings, Washington, DC, 1981.

Space

Surveillance is currently the most important military application of space. Satellites make it possible to watch enemy missile bases and thus to shorten the warning time. Satellites equipped with cameras move on an orbit that varies between 150 and 500 km high.

Conquest of Space
Largely dominated by the U.S. and the USSR

Year	1951–1960	1961–1965	1966–1970	1971–1975	1976–1980	1981	Total
USA	31	239	244	135	111	18	778
USSR	9	121	335	413	461	98	1,437

Table of successful space launches (1957-1981). 1982 (Jan.–April): U.S., 8; USSR, 29.

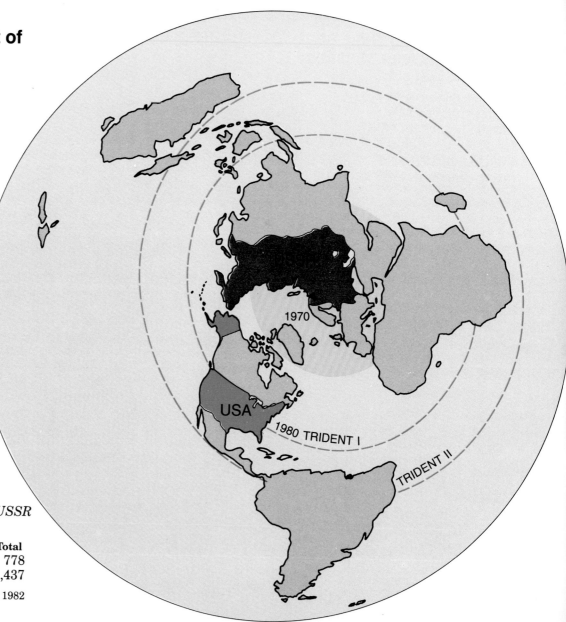

PRINCIPAL BENEFICIARIES OF AMERICAN MILITARY AID (1950-1980)
(In billions of dollars)

ESF

World total:	28
Israel	4
Egypt	4
S. Korea	2
Turkey	1
Jordan	0.9

FMS

World total:	22
Israel	12
Egypt	1.5
S. Korea	1.2
Turkey	1
Greece	1
Taiwan	0.5
Iran	0.5
Spain	0.5
Jordan	0.4

IMET

World total:	2
S. Vietnam	0.3
S. Korea	0.2
Turkey	0.1
Thailand	0.1

MAP

World total:	54
S. Vietnam	14.8
S. Korea	5.3
France	4
Turkey	3.1
Taiwan	2.6
Italy	2

ESF   Economic support fund
FMS   Foreign military sales financing program
IMET   International military education and training program
MAP   Military assistance program

FORCES FACE-TO-FACE IN THE EUROPEAN THEATER (1982)

NATO
(excluding Canada, including France)
American forces stationed in Europe
- Total strength: 0.22 million (0.20 million in FRG)
- Tanks: about 1,000
- Planes: more than 800
- Sixth Fleet (Mediterranean)
- Units of the Second Fleet (Holy Loch, GB)

European forces (NATO)
- Total strength: 2.2 million
- Tanks: about 16,000
- Fighter planes: about 3,700
- 260 major surface vessels, 143 submarines

WARSAW PACT
Soviet forces stationed in Central Europe*
- Total strength: 0.56 million (0.38 million in GDR)
- Tanks: about 10,000 (7,000 in Czech. and GDR)
- Fighter planes: about 2,000 (900 in GDR)
- Mediterranean Squadron (Black Sea Fleet)
- Baltic Fleet

East European forces (Warsaw Pact)
- Total strength: 0.85 million
- Tanks: about 14,500
- Fighter planes: about 2,250
- About 10 major surface vessels, 6 submarines

*Because of their proximity to Europe, it is necessary to add the Soviet forces stationed in the western part of the USSR and in the Caucasus region: 69 divisions (63 armored divisions).

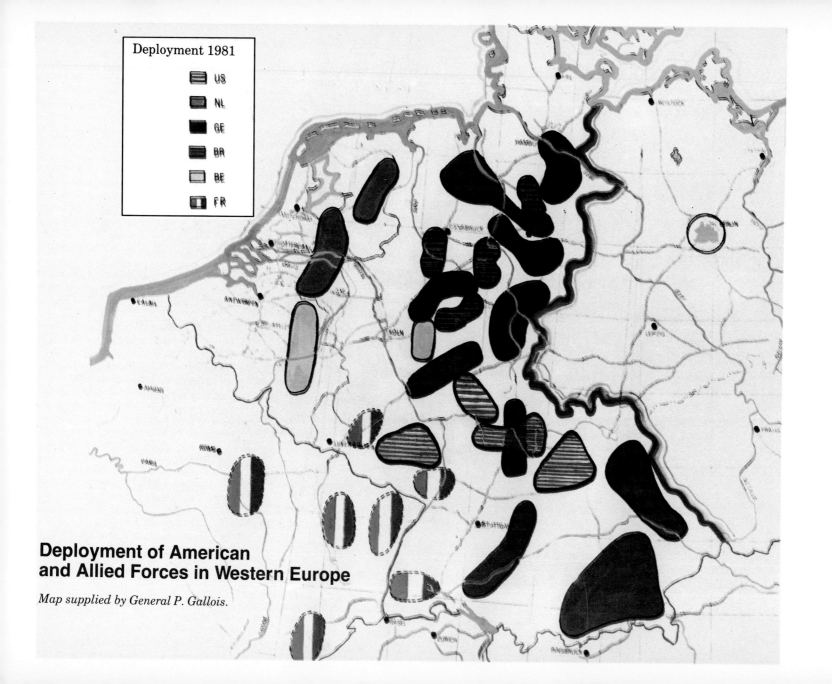

Deployment 1981

US
NL
GE
BR
BE
FR

**Deployment of American
and Allied Forces in Western Europe**

Map supplied by General P. Gallois.

Chronology: Major Weapons Systems

Year	1945	1950	1955	1960	1965	1970	1975	1980	1983
Politico-military events	Israel 48	Korea 50; NATO 49 52	End Vietnam I 54; Bandung 55	Hungary 56; Suez 56; Kennedy 60; Start Vietnam II 61	Cuba 62; France leaves NATO 66; Six-Day War 67	Czech. 68; Bangladesh 71; Paris Accords 73; Yom Kippur War 73	Fall of Saigon 75; Ogaden 77	Afghanistan 79-80; Vietnam China 79; Poland 80-81	Falklands 82; Lebanon 82
Nuclear events	Hiroshima Nagasaki 45	Atomic bomb USSR 49	Atomic bomb GB 52; H bomb US 52; H bomb USSR 53	H bomb GB 57; Atomic bomb Fr 60	Atomic bomb China 64; H bomb China 67	H bomb Fr 68	Atomic bomb India 74		

Weapons systems and vehicles

Year	1945	1950	1955	1960	1965	1970	1975	1980	1983
US — Strategic		(B 47 B 52 51 53)	1st nuclear sub US 54	1st satellite Explorer 58; Minute Man I 61; Titan II 62; 1st SLBM Polaris I 61	Minute Man II 66; Polaris III 64 MIRV	Moon landing 69; Minute Man III MIRV 70; Poseidon III 71 MIRV		Space shuttle 81; Trident IV 80	ACLM 82
US — Tactical		Honest John 53	Thor Jupiter 59-63	Sergeant 62; Pershing I 64	FB III 69; Lance 72				Pershing II 83; GLCM 83
USSR — Strategic		1st SLBM SSN4 55; Bear 56	SSN4 57-65	Sputnik 57; 1st ICBM SS7 61	SS9-SS11 65-66; SSN5 H 64; Yankee 67		SS17 (MIRV) 75; SS18-SS19 74 75		
USSR — Tactical		TV16 (Badger) 55; Scud-Frog 57	SS4 59	SS5 61			TU 22M (Backfire) 74	SS21-SS22-SS23 78 79 80; SS20 77	
Others			Buccaneer GB 62	Mirage IV Fr 64; 1st SLBM Fr.-67; Polaris III GB-67	1st missile Fr 68-70; Pluton Fr-73	M20 Fr-77		MRBM China-80; S3 Fr-80	

Source: Pierre Saint Macary, *Conflicts in the Contemporary World,* FNSP, 1982. Revised and expanded by Michel Tatu (1983).

American Aggressiveness as Seen by the USSR

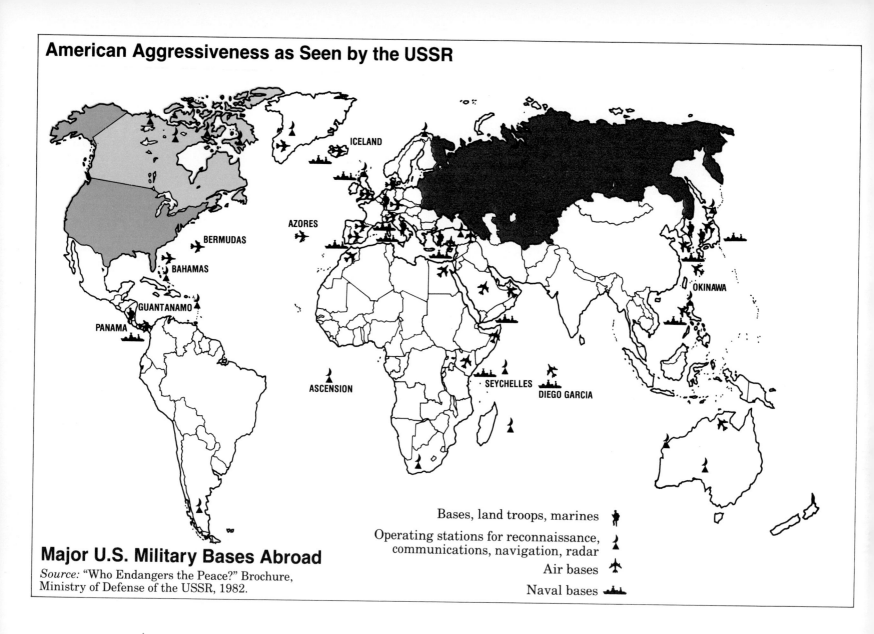

ICELAND

AZORES

BERMUDAS

BAHAMAS

GUANTANAMO

PANAMA

ASCENSION

SEYCHELLES

DIEGO GARCIA

OKINAWA

Bases, land troops, marines

Operating stations for reconnaissance, communications, navigation, radar

Air bases

Naval bases

Major U.S. Military Bases Abroad

Source: "Who Endangers the Peace?" Brochure, Ministry of Defense of the USSR, 1982.

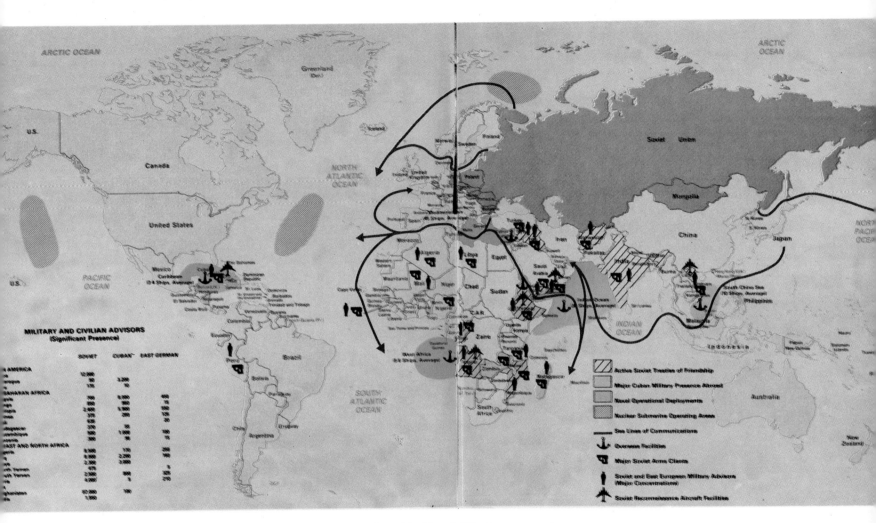

Soviet Aggressiveness as Seen by the United States

Source: C. W. Weinberger, *Soviet Military Power*, Dept. of Defense, Washington, DC, 1982.

Distances

km

km	Berlin	Bombay	Buenos Aires	Cairo	Calcutta	Caracas	Copenhagen	Chicago	Darwin	Hong Kong	Honolulu	Johannesburg	Lagos	London	Los Angeles	Lisbon	Mexico	Moscow	Nairobi	New York	Paris	Peking	Reykjavik	Rio de Janeiro	Rome	Singapore	Sydney	Tokyo
Berlin																												
Bombay	6288																											
Buenos Aires	11909	14925																										
Cairo	2890	4355	11814																									
Calcutta	7033	1664	16524	5699																								
Caracas	8435	14522	5096	10203	15464																							
Copenhagen	357	6422	12067	3206	7072	8392																						
Chicago	7084	12953	9011	9860	12839	4027	6840																					
Darwin	12946	7257	14693	11612	6047	18059	12903	15065																				
Hong Kong	8754	4317	18478	8150	2659	16360	8671	12526	4271																			
Honolulu	11764	12914	12164	14223	11343	9670	11407	6836	8640	8921																		
Johannesburg	8870	6974	8088	6267	8459	11019	9225	13984	10639	10732	19206																	
Lagos	5198	7612	7916	3915	9216	7741	5530	9612	14222	11854	16308	4505																
London	928	7190	11131	3508	7961	7507	952	6356	13848	9623	11632	9071	5017															
Los Angeles	9311	14000	9852	12200	13120	5812	9003	2804	12695	11639	4117	16676	12414	8758														
Lisbon	2311	8018	9600	3794	9075	6509	2478	6424	15114	11028	12587	8191	3799	1588	9122													
Mexico	9732	15656	7389	12372	15280	3586	9514	2726	14631	14122	6085	14585	11071	8936	2493	8676												
Moscow	1610	5031	13477	2902	5534	9938	1561	8000	11350	7144	11323	9161	6254	2498	9769	3906	10724											
Nairobi	6370	4532	10402	3536	6179	11544	6706	12883	10415	8776	17282	2927	3807	6819	15544	6461	14818	6244										
New York	6385	12541	8526	9020	12747	3430	6188	1145	16047	12950	7980	12841	8477	5572	3936	5422	3364	7510	11842									
Paris	876	7010	11051	3210	7858	7625	1026	6650	13812	9630	11968	8732	4714	342	9085	1454	9200	2486	6485	5836								
Peking	7822	4757	19268	7544	3269	14399	7202	10603	6011	1963	8160	11710	11457	8138	10060	9668	12460	5794	9216	10988	8217							
Reykjavik	2385	8335	11437	5266	8687	6915	2103	4757	13892	9681	9787	10938	6718	1887	6936	2848	7460	3304	8683	4206	2228	7882						
Rio de Janeiro	10025	13409	1953	9896	15073	4546	10211	8547	16011	17704	13342	7113	6035	9299	10155	7734	7693	11562	8928	7777	9187	17338	9874					
Rome	1180	6175	11151	2133	7219	8363	1531	7739	13265	9284	12916	7743	4039	1431	10188	1861	10243	2376	5391	6888	1105	8126	3297	9214				
Singapore	9944	3914	15879	8267	2897	18359	9969	15078	3349	2599	10816	8660	11145	10852	14123	11886	16610	8428	7460	15339	10737	4478	11514	15712	10025			
Sydney	16096	10160	11800	14418	9138	15343	16042	14875	3150	7374	8168	11040	15519	16992	12073	18178	12696	14497	12153	15989	16962	8949	16617	13501	16324	6300		
Tokyo	8924	6742	18362	9671	5141	14164	8696	10137	5431	2874	6202	13547	13480	9562	8811	11149	11304	7485	11260	10849	9718	2099	8802	18589	9861	5321	7823	
Vancouver	7980	12267	11294	10838	11445	6701	7646	2851	12215	10246	4357	16452	11951	7582	1736	8287	3945	8203	14531	3903	7923	8516	5698	11238	8991	12829	12501	7554

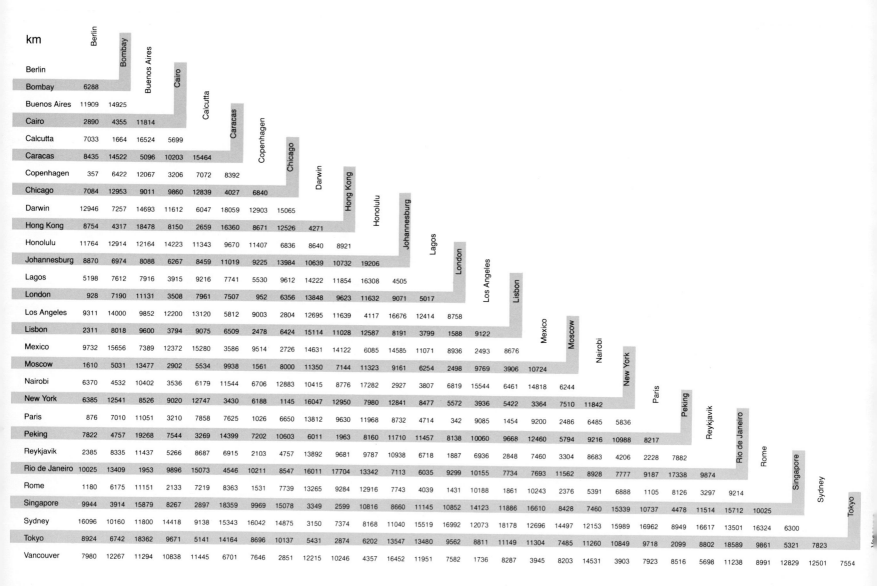

World Statistical Survey

	Land area (thousands of sq. km.)	Population, 1980 (in millions)	Density (pop./sq. km.)	Estimation of population, 2000 (in millions)	Urban population, 1980 (% total)	Life expectancy (in years)	% Adult literacy	Annual consumption of energy per person (in kilograms equiv. coal)	Contribution in calories per day & per person (1977)	GNP 1980 per person (in dollars)
	1	2	3	4	5	6	7	8	9	10
NORTH AMERICA										
Canada	9976	24	2	28	80	74	99	13200	3374	10310
United States	9363	228	24	259	77	73	99	11700	3576	11360
EUROPE										
Albania	29	2.7	93	4	37	70		1100	2730	(840)
E. Germany	108	17	155	17	76	73	99	7130	3641	7180
W. Germany	249	61	248	62	85	72	99	6260	3381	13590
Austria	84	7.5	90	8	54	72	99	5100	3535	10230
Belgium	31	9.8	321	10	72	73	99	6500	3583	12180
Bulgaria	111	9	81	10	64	73	98	5500	3611	4150
Cyprus	9	0.66	67		(43)	73	89	(1966)		3560
Denmark	43	5.1	119	5	84	75	99	5700	3418	12950
Eire	70	3.3	46	4	58	73	98	3700	3541	4880
Spain	505	37.5	74	43	74	73		2700	3149	5400
Finland	337	4.9	14	5	62	73	100	6000	3100	9720
France	548	54	98	58	78	74	99	4800	3434	11730
Greece	132	9.6	72	11	62	74		2160	3400	4380
Hungary	93	10.8	115	11	54	71	99	3800	3521	4180
Iceland	103	0.23	2		(88)	76	99			11330
Italy	301	57	189	61	69	73	98	3300	3428	6480
Luxembourg	3	0.36	138		(68)	72	99	(14700)	(3410)	14510
Malta	0.3	0.36	1070			72		(1080)		3470
Norway	324	4.1	13	4	53	75	99	11750	3175	12650
Netherlands	41	14.1	343	16	76	75	99	6600	3338	11470
Poland	313	35.8	115	42	57	71	98	5750	3656	3900
Portugal	92	9.9	106	11	31	71	72	1450	3076	2370
Romania	238	22.2	92	25	50	71	98	4660	3444	2340

	1	2	3	4	5	6	7	8	9	10
United Kingdom	245	56	229	58	91	73	99	5280	3336	7920
Sweden	450	8.3	18	8	87	75	99	8260	3221	13520
Switzerland	41	6.5	153	7	58	75	99	5000	3485	16440
Czechoslovakia	128	15.3	118	17	63	71	99	6660	3430	5820
USSR	22400	265	12	312	62	70		5800	3460	4550
Yugoslavia	256	22.4	87	26	42	69	85	2420	3445	2620
ASIA										
Afghanistan	648	15.9	24	24	15	37	12	88	2695	(170)
Saudi Arabia	2150	9	4	15	67	54	16	2000	2624	11260
Bahrain	0.6	0.4	620		(80)	67	(40)	(10000)		5560
Bangladesh	144	90	625	141	11	46	26	40	2100	130
Burma	677	35	48	54	27	54	70	67	2286	170
Bhutan	47	1.3	27	2	4	44			2028	80
China	9597	980	98	1245	13	64	66	734	2441	290
North Korea	121	18.3	151	28	60	65		2775	2837	(1130)
South Korea	98	38.8	393	52	55	65	93	1473	2785	1520
U. A. Emirates	84	1	11	1	72	63	56	4450		26850
Hong Kong	1	5.1	4900	6	90	74	90	1480	2883	4240
India	3288	680	207	994	22	52	36	194	2021	240
Indonesia	1919	148	77	216	20	53	62	225	2272	430
Iraq	435	13.1	30	23	72	56		664	2134	3020
Iran	1648	38.8	23	61	50	59	50	1141	3138	
Israel	21	3.9	178	5	89	72		3510	3141	4500
Japan	372	117	314	130	78	76	99	4050	2949	9890
Jordan	98	3.2	31	6	56	61	70	522	2107	1420
Cambodia	181	6.9	38	10	(15)			2	1926	
Kuwait	18	1.36	73	2	88	70	60	6160		19830
Laos	237	3.4	14	5	14	43	41	98	2082	
Lebanon	10	2.7	270	4	76	66		1028	2495	
Malaysia	330	14	42	21	29	64		713	2610	1620
Maldive Is.	0.3	0.15	473			47	82			260
Mongolia	1565	1.7	1	3	51	64		1483	2523	(780)
Nepal	141	14.6	96	22	5	44	19	13	2002	140
Oman	300	0.9	4			48		(2006)		4380
Pakistan	804	83	103	134	28	50	24	210	2280	300
Philippines	300	49	163	77	36	64	75	330	2189	690
Qatar	11	0.2	18			58				26080
Singapore	0.5	2.4	4137	3	100	72		5780	3074	4430

	1	2	3	4	5	6	7	8	9	10
Sri Lanka	66	15	220	21	27	66	85	135	2126	270
Syria	185	9	44	16	50	65	58	925	2684	1340
Taiwan	36	18	500	25	46					
Thailand	514	47	91	68	14	63	84	353	1929	670
Turkey	781	45	56	67	47	62	60	770	2907	1470
Vietnam	330	55	17	88	19	63	87	138	1801	?
Yemen	195	7	35	11	10	42	21	58	2192	430
S. Yemen	333	2	6	3	37	45	40	510	1945	420
AFRICA										
South Africa	1221	30	24	52	50	61		2900	2831	2300
Algeria	2382	19	8	34	44	56	35	645	2372	1870
Angola	1247	7.1	5	12	21	42		200	2133	470
Benin	113	3.4	30	6	14	47	25	65	2249	310
Botswana	600	0.8	1	1.3	(12)	50	(35)			910
Burundi	28	4.1	153	7	2	42	23	17	2254	200
Cameroon	475	8.4	17	14	35	47		143	2069	670
Cape Verde Is.	4	0.32	78		(20)	61		(143)		300
Central African Rep.	623	2.3	3	4	41	44	39	46	2242	300
Comoros Is.	2	0.34	147			(50)		(52)		300
Congo	342	1.6	4	3	45	59		195	2284	900
Ivory Coast	322	8.4	25	15	40	47	41	230	2517	1150
Djibouti	22	0.4	18			45	14	(823)		480
Egypt	1001	41	40	60	45	57	44	540	2760	580
Ethiopia	1222	31	24	54	14	40	15	20	1754	140
Gabon	268	0.55	2		(32)	45		(1830)		4400
Ghana	239	12	46	23	36	49		260	1983	420
Guinea	246	5.5	19	9	19	45	20	83	1943	290
Guinea-Bissau	36	0.57	15			42	28	(62)		160
Equat. Guinea	28	0.36	12			47		(105)		
Upper Volta	274	6.1	24	10	10	39	5	26	1875	210
Kenya	583	16	26	36	14	55	50	172	2032	420
Lesotho	30	1.3	43	2	12	51	52		2245	420
Liberia	111	1.9	16	4	28	54	25	425	2404	530
Libya	1760	3	2	5	30	56	50	2250	2985	8640
Madagascar	587	8.7	14	16	18	47	50	90	2486	350
Malawi	118	6.1	48	12	10	44	25	67	2066	230
Mali	1240	7	5	13	17	43	9	28	2117	190
Morocco	447	20	42	36	40	56	28	302	2534	900

	1	2	3	4	5	6	7	8	9	10
Mauritius	2	1	452	(1,8)	(44)	65	85	(407)		1060
Mauritania	1031	1.5	1	3	23	43	17	196	1976	440
Mozambique	783	12	13	22	9	47	28	120	1906	230
Namibia	825	1.01	1	(1,8)						
Niger	1267	5.3	4	10	13	43	5	46	2139	330
Nigeria	924	85	90	169	20	49	30	80	1951	1010
Uganda	236	13	54	24	9	54	48	39	2110	300
Rwanda	26	5.2	171	10	4	45	50	28	2264	200
São Tomé & Principe	1	0.1	103					(173)		490
Senegal	197	5.7	27	10	32	43	10	253	2261	450
Seychelles Is.	0.3	0.1	220		(27)	66		(583)		1770
Sierra Leone	72	3.5	46	6	22	45		84	2150	280
Somalia	638	4	6	7	30	44	60	74	2033	
Sudan	2506	19	7	34	25	46	20	133	2184	410
Swaziland	17	0.55	33	1	(8)	47	65			680
Tanzania	945	19	20	36	12	52	66	51	2063	280
Chad	1284	4.1	3	7	18	41	15	22	1762	120
Togo	57	2.5	43	5	20	47	18	112	2069	410
Tunisia	164	6.5	37	10	52	60	62	590	2674	1310
Zaire	2345	29	12	51	34	47	58	100	2271	220
Zambia	753	5.8	7	11	43	49	44	832	2002	560
Zimbabwe	391	7.4	18	17	23	55	74	783	2576	630
LATIN AMERICA										
Argentina	2767	28	10	34	82	70	93	1960	3347	2390
Bahamas	14	0.24	16		(58)	69	93	(7350)		3790
Belize	23	0.16	7					(598)		1080
Bolivia	1099	5.6	5	9	33	50	63	447	1974	570
Brazil	8512	119	14	177	68	63	76	1020	2562	2050
Chile	757	11	14	15	80	67		1153	2656	2150
Colombia	1139	27	23	39	70	63		914	2364	1180
Costa Rica	51	2.2	41	3	43	70	90	812	2550	1730
Cuba	115	9.8	83	12	65	73	96	1360	2720	(1410)
Dominique	0.5	0.08	170					(203)		620
El Salvador	21	4.5	206	8	41	63	62	338	2051	660
Ecuador	284	8	28	14	45	61	81	640	2104	1270

	1	2	3	4	5	6	7	8	9	10
Grenada	0.8	0.1	130			69		(209)		690
Guatemala	109	7.3	61	12	39	59		(260)	2156	1080
Guyana	215	0.9	4	1.5	(46)	70		(1070)		690
Haiti	28	5	174	7	28	53	23	63	2100	270
Honduras	112	3.7	31	7	36	58	60	238	2015	560
Jamaica	11	2.2	195	3	41	71	90	1326	2660	1040
Barbados	0.4	0.25	620		(46)	71	99	(1100)		3040
Mexico	1973	70	36	115	67	65	81	1535	2654	2090
Nicaragua	130	2.6	18	5	53	56	90	446	2446	740
Panama	77	1.8	25	3	54	70		895	2341	1730
Paraguay	407	3.2	7	5	39	65	84	234	2824	1300
Peru	1285	18	13	27	67	58	80	716	2274	930
Dominican Rep.	49	5.5	112	9	51	61	67	156	2094	1160
St. Lucia	0.7	0.12	195					(392)		900
St. Vincent	0.4	0.12	300					(211)		520
Surinam	163	0.4	2			68	65	(2150)		2840
Trinidad & Tobago	5	1.2	240	2	(23)	72	95	4870	2694	4370
Uruguay	176	2.9	16	4	84	71	94	1220	3036	2810
Venezuela	912	15	14	24	83	67	82	2950	2435	3630
AUSTRALIA AND THE PACIFIC										
Australia	7687	14.5	2	17	89	74	100	6540	3428	9820
New Zealand	269	3.3	12	4	85	73	99	4700	3345	7090
Papua-New Guinea	462	3	6	5	18	51	32	300	2268	780
Fiji	18	0.63	33		(37)	75	65	(466)		1850
Kiribati	1	0.07	71		(30)			(317)		
Nauru	0.02	0.008	380					(7000)		
Solomon	28	0.22	8		(10)			(200)		480
Western Samoa	3	0.16	56		(21)	68		(187)		
Tonga	0.7	0.1	142					(169)		
Tuvalu										
Vanuatu (New Hebrides)	15	0.11	7		(24)			(480)		530

Source: Report on World Development, 1982, World Bank.
Note: Earlier sources or other sources in parentheses.

Bibliographical References

Report on World Development. World Bank. Washington, DC, annual.

Ramsès. *Annual Report of the French Institute of International Relations.* Paris, annual.

World Economic Outlook. International Monetary Fund. Washington, DC, annual.

Economic Perspectives. OECD. Paris, annual.

Strategic Survey. International Institute for Strategic Studies. London, annual.

The Military Balance. International Institute for Strategic Studies. London, annual.

SIPRI Yearbook. Stockholm.

Specialized publications of the Documentation française. Paris.

Specialized publications of the U.S. Department of State. Washington, DC.

Among the French journals: *Politique étrangère, Politique internationale, Commentaire, Défense nationale, Stratégique, Études polémologiques, Géopolitique. Hérodote* (the only journal of geography that integrates the geopolitical dimension).

Among the English-language journals: *Foreign Affairs, Foreign Policy, The Washington Quarterly.*